STUDENT UNIT GUIDE

AS Human Biology
UNIT 3

Specification A

Module 3: Pathogens and Disease

Steve Potter

AS Human Biology

Philip Allan Updates
Market Place
Deddington
Oxfordshire
OX15 0SE

Tel: 01869 338652
Fax: 01869 337590
e-mail: sales@philipallan.co.uk
www.philipallan.co.uk

© Philip Allan Updates 2004

ISBN 0 86003 939 0

All rights reserved; no part of this publication may be reproduced, stored in a retrieval system, or transmitted, in any form or by any means, electronic, mechanical, photocopying, recording or otherwise without either the prior written permission of Philip Allan Updates or a licence permitting restricted copying in the United Kingdom issued by the Copyright Licensing Agency Ltd, 90 Tottenham Court Road, London W1P 9HE.

This Guide has been written specifically to support students preparing for the AQA Specification A AS Human Biology Unit 3 examination. The content has been neither approved nor endorsed by AQA and remains the sole responsibility of the author.

Printed by Information Press, Eynsham, Oxford

Contents

Introduction
About this guide 4
Preparing for the Unit 3 test 5
Approaching the unit test 6

■ ■ ■

Content Guidance
About this section 10
Disease 11
Parasites and parasitism 22
Defence against infection 26
Non-communicable diseases 32
The cell cycle, mitosis and meiosis 37
The structure and functions of DNA and RNA 42
Genetic engineering 49
Diagnosis and treatment of disease 54

■ ■ ■

Questions and Answers
About this section 64
Q1 The structure of DNA and RNA 65
Q2 Bacterial growth 67
Q3 Mitosis and meiosis 70
Q4 Protein synthesis 73
Q5 Immunity 75
Q6 Tumours 77
Q7 Malaria 79
Q8 Analysing gene structure 81
Q9 Heart disease 84
Q10 Tuberculosis 87

AS Human Biology

Introduction

About this guide

This guide is written to help you to prepare for the Unit 3 examination of the AQA Human Biology Specification A. Unit 3 examines the content of **Module 3: Pathogens and Disease**, and forms part of the AS assessment. It will also form part of the A2 assessment with some of the material being re-examined at the end of the A2 synoptic examination.

This Introduction provides guidance on revision, together with advice on approaching the examination itself.

The Content Guidance section gives a point-by-point description of all the facts you need to know and concepts you need to understand for Module 3. Although each fact and concept is explained where necessary, you must be prepared to use other resources in your preparation.

The Question and Answer section shows you the sort of questions you can expect in the unit test. It would be impossible to give examples of every kind of question in one book, but these should give you a flavour of what to expect. Each question has been attempted by two candidates, Candidate A and Candidate B. Their answers, along with the examiner's comments, should help you to see what you need to do to score a good mark — and how you can easily *not* score a mark even though you probably understand the biology.

What can I assume about the guide?

You can assume that:
- the topics described in the Content Guidance section correspond to those in the specification
- the basic facts you need to know are stated clearly
- the major concepts you need to understand are explained
- the questions at the end of the guide are similar in style to those that will appear in the unit test
- the answers supplied are genuine answers — not concocted by the author
- the standard of the marking is broadly equivalent to the standard that will be applied to your answers

What can I not assume about the guide?

You *must not* assume that:
- every last detail has been covered
- the diagrams used will be the same as those used in a unit test (they may be more or less detailed, seen from a different angle etc.)
- the way in which the concepts are explained is the *only* way in which they can be presented in an examination (often concepts are presented in an unfamiliar situation)

- the range of question types presented is exhaustive (examiners are always thinking of new ways to test a topic)

So how should I use this guide?

The guide lends itself to a number of uses throughout your course — it is not *just* a revision aid. Because the Content Guidance is laid out in sections that correspond to those of the specification for Module 3, you can:
- use it to check that your notes cover the material required by the specification
- use it to identify strengths and weaknesses
- use it as a reference for homework and internal tests
- use it during your revision to prepare 'bite-sized' chunks of related material, rather than being faced with a file full of notes

The Question and Answer section can be used to:
- identify the terms used by examiners in questions and what they expect of you
- familiarise yourself with the style of questions you can expect
- identify the ways in which marks are lost as well as how they are gained

Preparing for the Unit 3 test

Preparation for examinations is a very personal thing. Different people prepare, equally successfully, in very different ways. The key is being totally honest about what actually *works* for *you*. This is *not* necessarily the same as the style you would like to adopt. It is no use preparing to a background of rock music if this distracts you.

Whatever your style, you must have a plan. Sitting down the night before the examination with a file full of notes and a textbook does not constitute a revision plan — it is just desperation — and you must not expect a great deal from it. Whatever your personal style, there are a number of things you *must* do and a number of other things you *could* do.

Things you must do

- Leave yourself enough time to cover all the material.
- Make sure that you actually have all the material to hand (use this book as a basis).
- Identify weaknesses early in your preparation so that you have time to do something about them.
- Familiarise yourself with the terminology used in examination questions (see p. 6).

Things you could do to help you learn

- Copy selected portions of your notes.
- Write a precis of your notes which includes all the key points.
- Write key points on postcards (carry them round with you for a quick revise during a coffee break).

- Discuss a topic with a friend also studying the same course.
- Try to explain a topic to someone not on the course.
- Practise examination questions on the topic.

Approaching the unit test

Terms used in examination questions

You will be asked precise questions in the examinations, so you can save a lot of valuable time as well as ensuring you score as many marks as possible by knowing what is expected. Terms most commonly used are explained below.

Describe
This means exactly what it says — 'tell me about...' — and you should not need to explain why.

Explain
Here you must give biological reasons for *why* or *how* something is happening.

Complete
You must finish off a diagram, graph, flow chart or table.

Draw/plot
This means that you must construct some type of graph. For this, make sure that:
- you choose a scale that makes good use of the graph paper (if a scale is not given) and does not leave all the plots tucked away in one corner
- plot an appropriate type of graph — if both variables are continuous variables, then a line graph is usually the most appropriate; if one is a discrete variable, then a bar chart is appropriate
- plot carefully using a sharp pencil and draw lines accurately

From the...
This means that you must use only information in the diagram/graph/photograph or other forms of data.

Name
This asks you to give the name of a structure/molecule/organism etc.

Suggest
This means 'give a plausible biological explanation for' — it is often used when testing understanding of concepts in an unfamiliar situation.

Compare
In this case you have to give similarities *and* differences between...

Calculate
This means add, subtract, multiply, divide (do some kind of sum!) and show how you got your answer — *always show your working!*

When you finally open the test paper, it can be quite a stressful moment. You may not recognise the diagram or graph used in question 1. It can be quite demoralising to attempt a question at the start of an examination if you are not feeling very confident about it. So:
- *do not* begin to write as soon as you open the paper
- *do not* answer question 1 first, just because it is printed first (the examiner did not sequence the questions with your particular favourites in mind)
- *do* scan *all* the questions before you begin to answer any
- *do* identify those questions about which you feel most confident
- *do answer first* those questions about which you feel most confident regardless of order in the paper
- *do read the question carefully* — if you are asked to explain, then explain, don't just describe
- *do* take notice of the mark allocation and don't supply the examiner with all your knowledge of osmosis if there is only 1 mark allocated (similarly, you will have to come up with four ideas if 4 marks are allocated)
- *do* try to stick to the point in your answer (it is easy to stray into related areas that will not score marks and will use up valuable time)
- *do* take care with
  - drawings — you will not be asked to produce complex diagrams, but those you do produce must resemble the subject
  - labelling — label lines *must touch* the part you are required to identify; if they stop short or pass through the part, you will lose marks
  - graphs — draw *small* points if you are asked to plot a graph and join the plots with ruled lines or, if specifically asked for, a line or smooth curve of best fit through all the plots
- *do try* to answer *all* the questions

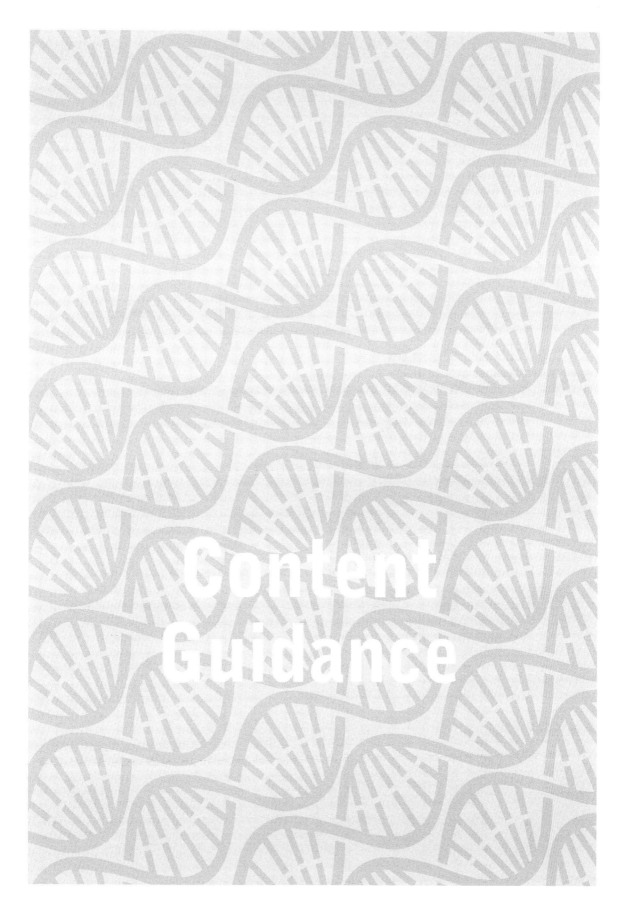

This section is a guide to the content of **Module 3: Pathogens and Disease**. The main areas of this module are:
- Disease
- Parasites and parasitism
- Defence against infection
- Non-communicable diseases
- The cell cycle, mitosis and meiosis
- The structure and functions of DNA and RNA
- Genetic engineering
- Diagnosis and treatment of disease

You should think of this section as a 'translation' of the specification from 'examiner speak' into more user-friendly language. At the same time I have tried to be very precise in describing exactly what is required of you.

Key facts you must know
These are exactly what you might think: a summary of all the basic knowledge that you must be able to recall. All the actual knowledge has been broken down into a number of small facts that you must learn. This means that the list of 'Key facts' for some topics is quite long. However, this approach makes quite clear *everything* you need to know about the topic.

Key concepts you must understand
Whereas you can learn facts, you must *understand* these ideas or concepts. You can know the actual words that describe a concept like mitosis, or DNA replication, but you will not be able to use this information unless you really understand what is going on. I have given brief explanations of all the major concepts, but you must be prepared to refer to your notes and textbooks or ask your teacher for a fuller explanation.

What the examiners will expect you to be able to do
In this part, I have tried to give you an insight into the minds of the examiners who will set and mark your examination papers. Obviously, they may ask you to recall any of the basic knowledge or explain any of the key concepts; but they may well do more than that. Examiners think up questions where the concepts you understand are in a different setting or context from the one(s) you are familiar with. I have tried, in this section, to prepare you for the sorts of questions they might ask. This can never be exhaustive, but it will give you a good idea of what can be asked of you. Bear in mind that examiners will often set individual questions that involve knowledge and understanding of more than one section. The sample questions in the Question and Answer section of this book will help you to practise this skill.

After each topic there is a short paragraph marked 'Links'. While not crucial to the understanding of any of the biology, this should give you some idea of how the biology you are learning will be related to other topics that you may meet in other modules.

Disease

Definitions

Key concepts you must understand

Disease is a condition with a specific cause in which part or all of an organism is made to function in an abnormal or less efficient manner. Organisms that cause disease are called **pathogens**. The process by which a pathogen enters and becomes established in an organism to cause disease is called **infection**.

Tip Don't refer to the disease itself as an infection. Disease is a condition; infection is a process.

The causes of disease include:
- pathogenic organisms (bacteria, viruses, fungi and protoctistans). Diseases caused in this way are known as **infectious diseases**. Infectious diseases that can be transferred from one person to another are called **communicable diseases**.
- a person's lifestyle and working conditions. These may result in **human-induced diseases**. Examples include many cancers, some forms of heart disease and asbestosis.
- degenerative processes. These are often the result of ageing. Arthritis and atherosclerosis are examples of **degenerative diseases**.
- genetics. Haemophilia and phenylketonuria are examples of **genetic diseases**.
- nutrient deficiency. **Deficiency diseases** include scurvy (caused by a lack of vitamin C in the diet) and kwashiorkor (caused by a lack of protein in the diet).

Key facts you must know

Infection can take place in a number of ways. At any point where the body has contact with the environment, there is the potential for microorganisms to invade the body. The main methods of infection are described in the table below.

Method of infection	Notes on infection	Diseases spread in this way
Droplet infection	Many cause **respiratory diseases** affecting the bronchi and bronchioles — microorganisms are carried in tiny droplets when an infected person sneezes; another person then breathes them in	Common cold, influenza, pneumonia, TB
Drinking contaminated water	Microorganisms in the water infect cells lining the gut and reproduce — they are then released back into the gut and pass out with the faeces	Cholera, typhoid fever
Eating contaminated food	Cells lining the gut are infected — microorganisms reproduce and are passed out with the faeces	Salmonellosis, botulism, listeriosis

AS Human Biology

Method of infection	Notes on infection	Diseases spread in this way
Direct contact	Skin infections are often transmitted in this way — contact with an infected person's skin or with a contaminated surface may result in the microorganism being transmitted to another person	Athlete's foot, ringworm
Sexual intercourse	Microorganisms infecting the sex organs may be transmitted during intercourse	Syphilis, AIDS, gonorrhoea
Blood to blood	Many sexually transmitted diseases can also be transmitted in this way — drug users sharing needles run the risk of such infections	AIDS, hepatitis B
Animal vectors	Many diseases are spread through insect bites — the female *Anopheles* mosquito transmits the malarial parasite when she sucks blood; in a similar way, tsetse flies transmit the organism that causes sleeping sickness	Malaria, sleeping sickness, typhoid fever, salmonellosis

What the examiners will expect you to be able to do
- Recall any of the key facts.
- Explain any of the key concepts.
- Relate methods of infection to immune responses by the body.

Pathogenic microorganisms

Key facts you must know

Most infectious diseases in humans are caused by either bacteria or viruses. Bacteria are prokaryotic cells; viruses are acellular (not made of cells). The diagrams below show their typical structures.

Structure of a typical bacterium

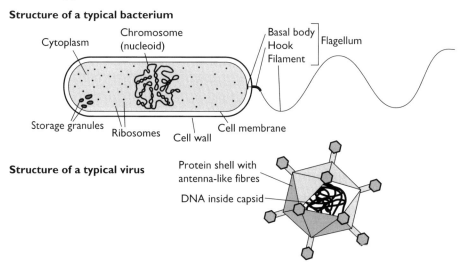

Structure of a typical virus

AQA (A) Unit 3

Key concepts you must understand

Given a supply of water and nutrients, together with the correct amount of oxygen, all bacteria can reproduce outside other living cells. Most bacteria are not pathogenic. However, viruses can only reproduce inside living cells, so *all* viruses are parasitic.

Bacteria usually reproduce by a form of **binary fission**. The genetic material replicates itself and then the bacterial cell divides into two. Following growth, the process is repeated. Under favourable conditions, reproduction can take place every 20 minutes. These conditions include:
- optimum temperature for enzyme activity in the bacterial cell
- optimum pH for enzyme activity in the bacterial cell
- a supply of water — a bacterial cell is about 80% water and so cannot be active without it
- a supply of nutrients to provide the 'raw materials' for synthesis of cellular molecular components and to act as a respiratory substrate
- in aerobic bacteria, an adequate supply of oxygen to allow aerobic respiration to release the energy necessary for reactions to proceed

Tip You may be expected to use knowledge from Module 1 to explain the effects on enzymes of pH and temperatures that are significantly different from the optimum values. Make sure that you can do this. Try to use the phrase *release energy* when describing the role of respiration. Energy can never be made/created and it can never be destroyed.

Bacterial growth curve

The progress of bacterial reproduction can be represented by a **bacterial growth curve** — a graph of the number of bacteria against time. ('Growth' refers to growth of the population of bacterial cells.)

The graph below shows a bacterial growth curve produced when a small number of bacteria are transferred to a new **liquid-culture medium** supplying all the necessary nutrients, either suspended or dissolved in water. The medium is usually aerated and adjusted to the optimum pH for the bacteria.

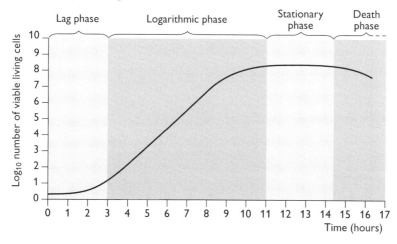

AS Human Biology

Tip Note that the y-axis has a base-10 logarithmic scale. On such a scale, each number represents 10 to the power of that number. Therefore, 1 represents 10^1 (10), 2 represents 10^2 (100), 3 represents 10^3 (1000) and 8 represents 10^8 (100 000 000). So the graph covers a wide range of bacterial cell numbers.

Lag phase
During the lag phase, there is little or no cell division. The bacterial cells are adapting to the new conditions. They may have to synthesise different enzymes to utilise the nutrients present. Therefore, different genes must be activated. The lag phase is longer when the new conditions are very different from the ones to which the bacteria had previously been exposed.

Logarithmic (log) phase
This is also called the **exponential phase**. Bacteria are reproducing quickly because nutrients, water and oxygen are readily available and the pH and temperature values are at or near the optimum. As a consequence, bacterial numbers are doubling at a constant, rapid rate.

Stationary phase
The numbers remain constant. The rapid reproduction of the log phase uses up nutrients and, as these become depleted, reproduction slows. Also, many of the cells die because of the accumulation of excretory products in the medium. The number of new cells formed is generally equal to the number of cells dying.

Death phase
As nutrients become even more depleted and the concentration of excretory products increases further, reproduction slows even more and more cells die. The number of cells dying is greater than the number of new cells being formed.

Tip You may be asked to sketch on a growth curve the effect of changing a variable, such as temperature. Remember, anything that increases the rate of reproduction will probably result in higher numbers. Therefore, nutrients will be used up more quickly and the population will enter the death phase sooner. The reverse is true for anything that slows reproduction.

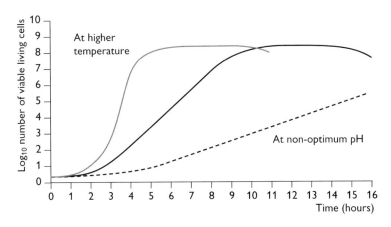

Practical techniques

Growing bacteria on a solid culture medium

This is usually an **agar**-based medium. Agar is a kind of gel. Adding a different mix of nutrients to the agar produces a specific **nutrient agar**. Different bacteria have different nutritional requirements. Therefore, by choosing a particular medium, a pure culture of one species can be obtained because this species will reproduce in the specific medium and others will not.

The agar is usually contained in a **Petri dish** and the bacteria to be cultured are introduced using an **inoculating loop**. As the bacteria reproduce in the agar, they form **colonies**, each of which contains many millions of bacteria.

It is essential to use **aseptic technique** when culturing bacteria. This minimises the chances of introducing any microorganisms into the medium other than those to be cultured. The following procedures are usually carried out when culturing bacteria on a solid medium:

- Bottles of agar are sterilised in an **autoclave**. An autoclave produces steam under high pressure, raising the temperature to between 115°C and 140°C. This ensures that there are no living microorganisms in the agar to be used. The agar is poured into sterile Petri dishes and allowed to set.

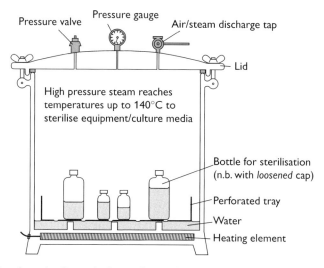

- The inoculating loop is 'flamed' (heated to red heat) before being used to transfer the **inoculum** of bacteria. This ensures that there are no living microorganisms on the loop, except those in the culture.
- When transferring the inoculum to the surface of the agar, the lid of the Petri dish is lifted only slightly. This minimises the risk of microorganisms entering from the atmosphere.
- The inoculating loop is rubbed gently over the surface of the agar, taking care not to 'dig in'.

- Once the inoculum has been transferred, the lid is replaced on the Petri dish and taped down. The Petri dish is then **incubated** at a temperature appropriate to the species being cultured. Agar plates are usually incubated upside down.

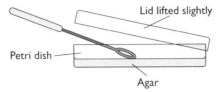

Tip The optimum temperature for most bacteria is not 37°C (body temperature). Pathogenic bacteria are adapted to have an optimum temperature that is the same as body temperature. However, most free-living bacteria never encounter temperatures this high and have much lower optimum temperatures.

Growing bacteria in a liquid medium

Bacteria are often grown in liquid media. These **nutrient broths** are water-based and contain a mix of nutrients. Different mixes of nutrients can be used to culture different types of bacteria. In addition, the pH of liquid media can be adjusted to the optimum for specific bacteria. The inoculating loop is flamed as before and the mouth of the vessel containing the medium is also passed through a Bunsen flame, to prevent microorganisms entering from the air.

Counting microorganisms in liquid cultures

The number of microorganisms in a liquid culture medium is estimated using a **haemocytometer**. This is a special type of microscope slide with a cavity of known depth and a grid precisely etched on the surface.

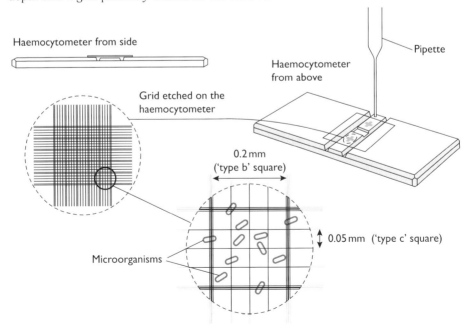

To use the haemocytometer, carry out the following procedure:
- Draw a small sample of a liquid culture into a Pasteur pipette.
- Rest the tip of the pipette on the central portion of the haemocytometer and against the coverslip. The culture will be drawn under the coverslip by capillary action.
- Allow 30 minutes for the microorganism cells to settle.
- Examine under a microscope and count the number of microorganisms in several of the 'type b' squares. Some cells will be 'half-in' and 'half-out'. Record half of these cells as being in the square.
- Obtain an average for the number of microorganisms per 'type b' square.
- Estimate the number of microorganisms in 1 cm^3 of culture in the following way:
  - Each 'type b' square has a side of 0.2 mm and therefore has an area of $0.2 \times 0.2 = 0.04$ mm^2.
  - The distance between the cover slip and the slide is exactly 0.1 mm, so the volume of culture in a 'type b' square is $0.04 \times 0.1 = 0.004$ mm^3.
  - 1 cm^3 = 1000 mm^3, so the volume in cm^3 of a 'type b' square is

  $$\frac{0.004}{1000} = 0.000004 \text{ cm}^3$$

  - If there were an average of 28 cells in a 'type b' square, the number of cells in 1 cm^3 would be:

  $$\frac{28 \times 1}{0.000004} = 7\,000\,000$$

Tip Bear in mind that the original culture may have been diluted to make counting easier. In this case you must multiply the answer obtained from the counting exercise by the dilution factor. For example, if the culture described above had been diluted by a factor of 20, then the actual number of cells in 1 cm^3 of the original culture would be:

$20 \times 7\,000\,000 = 140\,000\,000$

What the examiners will expect you to be able to do

- Recall any of the key facts.
- Explain any of the key concepts.
- Describe and explain any of the practical procedures.
- Describe differences between bacteria and viruses.
- Estimate the number of microorganisms in a culture from a drawing of a few squares of a haemocytometer grid, showing cells.
- Relate the structure of bacteria and viruses to the ways in which they cause disease.

The link between pathogens and infectious disease

Key concepts you must understand

Research in the nineteenth century established the link between microorganisms and infectious disease. In 1883, Robert Koch isolated the organism that causes cholera. He went on to list a number of criteria that must be met if an organism is to be identified as the cause of a specific disease. These are called **Koch's postulates**. They state that:

- the microorganism must *always* be present when the disease is present, and should *not* be present if the disease is not present
- the microorganism can be isolated from an infected person and then grown in culture
- introducing such cultured microorganisms into a healthy host should result in the disease developing
- it should then be possible to isolate the microorganism from this newly diseased host and grow it in culture

The first postulate establishes a link between the microorganism and the disease. The following three postulates prove that the metabolism of a specific living microorganism, when transferred into a healthy host, causes the disease.

Bacteria and viruses cause most infectious diseases in humans. The way in which they cause disease is linked to their size and structure. Bacteria are cellular and do not normally invade our cells following infection. However, their metabolism produces **toxins** that damage the body. Viruses invade our cells and, once inside, direct the cell's metabolism to produce more viruses. This results in cell death, which is the cause of the disease. For example, cell death in the mucous membranes lining the nose and throat when we have a cold results in these membranes becoming inflamed, leading to the typical runny nose and sore throat.

Key facts you must know

Specific diseases

Salmonellosis

Salmonellosis is a kind of food poisoning caused by bacteria in the genus *Salmonella*. One species (*Salmonella typhi*) causes typhoid fever. *Salmonella* bacteria can be transmitted in a number of ways, as shown in the diagram opposite. The main symptoms of the disease are:

- diarrhoea and vomiting
- fever
- headache and abdominal pain

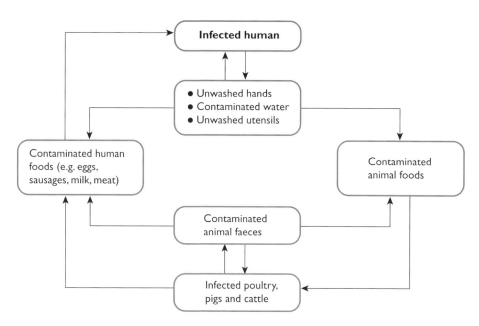

Diarrhoea and vomiting are the direct result of the effects of toxins released by the bacteria as they reproduce. *Salmonellae* are unusual in that they do enter the cells lining the intestines and multiply, releasing toxins inside the cells. This reduces the uptake of sodium ions and glucose from the small intestine. Water would normally follow the sodium ions and glucose by osmosis, as a water potential gradient is established between the lumen of the intestine and the epithelial cells. Without this water potential, water uptake is reduced and water may even be withdrawn from the epithelial cells by osmosis. This results in diarrhoea. Fever is one of the body's responses to infection. Raising the core temperature usually damages pathogens more than it does the bodies' own cells.

Pulmonary tuberculosis
Pulmonary tuberculosis (TB) is an infectious disease caused by the bacterium *Mycobacterium tuberculosis*. It is most commonly spread by droplet infection. The bacterium enters the lungs and multiplies, where it forms an 'infection focus'. The damage to the cells of the lungs creates cavities in the lung tissue, which can be detected by X-ray. Normally, the immune system halts the infection at this stage. However, it can spread into the lymph and infect other organs (e.g. the kidneys) as well as bones.

Although TB is an infectious disease, it is not just vaccination that has reduced the number of cases and deaths. Improvements in general hygiene have reduced the transmission of the disease and improvements in diet and living conditions have meant that people who contract the disease are much more likely to produce an effective immune response. The graph overleaf illustrates these points.

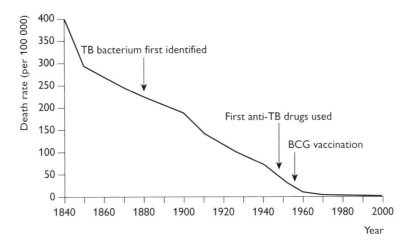

TB is treated with combinations of antibiotics. This reduces the risk of strains of the TB bacterium developing that are resistant to a particular antibiotic. Despite this, resistant strains are developing in some parts of the world.

Acquired immune deficiency syndrome (AIDS)

AIDS is caused by the **human immunodeficiency virus (HIV)**. The structure of HIV is shown below.

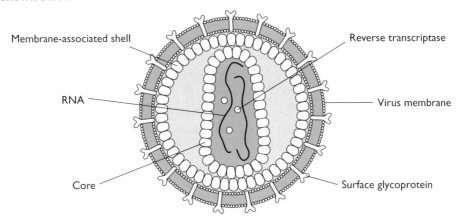

HIV is a **retrovirus** with RNA as the genetic material. Retroviruses stimulate host cells into converting their RNA into DNA, using an enzyme called **reverse transcriptase**. (See pages 51–52 for further details of this process. Reading the section on immune responses now will also help you to understand the significance of AIDS and the process of infection by HIV.)

Tip Remember, AIDS is the disease; HIV is the virus that causes the disease.

HIV infects the very cells that help defend the body against invading microorganisms — the **helper T-lymphocytes**. These cells stimulate other types of lymphocyte to produce their immune responses.

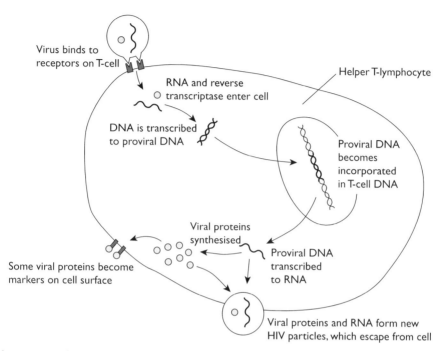

The course of infection of a helper T-lymphocyte by HIV occurs in three stages:
- **Binding** — glycoproteins on the surface of HIV bind to receptor sites on the surface of the helper T-lymphocytes. This allows the virus membrane to fuse with the plasma membrane of the T-lymphocyte. The viral RNA is released, together with the reverse transcriptase. This enzyme uses DNA nucleotides from the cell to make a DNA copy of the viral RNA. This **proviral DNA** now becomes integrated into the host cell's DNA.
- **Latency** — the proviral DNA remains dormant until the host-cell DNA is activated, either by **replication** of the DNA or by **transcription** of the DNA for protein synthesis.
- **Multiplication** — once the DNA is activated, the proviral DNA is also activated. It is transcribed to RNA, which:
  - initially directs the synthesis of viral proteins; these migrate to the cell surface and are displayed on the plasma membrane
  - becomes incorporated into new virus particles

During the multiplication phase, many of the infected cells are destroyed by:
- killer T-lymphocytes secreting chemicals that kill infected cells
- B-lymphocytes producing antibodies against HIV. The person is now **HIV-positive** — he/she would give a positive test for the presence of these antibodies.

The helper T-lymphocytes that are destroyed are replaced by others. These in turn become infected and are destroyed and replaced and so on. This cycle of destruction and replacement is called the **latency phase** of the disease (this is *not* the same as the latency stage of the viral DNA in any one helper T-cell).

The latency phase of the disease can last up to 20 years. However, eventually the body will fail to replace all the helper T-lymphocytes that have been destroyed. Then, the rest of the immune system — the B- and killer T-lymphocytes — will be less able to target *any* invading microorganisms and, as a result, diseases can develop more easily. These diseases are called **opportunistic infections** because they would not normally occur in a reasonably healthy person. Importantly, they include TB. AIDS is a major factor in the recent increase in the incidence of TB.

What the examiners will expect you to be able to do

- Recall any of the key facts.
- Explain any of the key concepts.
- Explain how infection by HIV ultimately makes the patient more susceptible to other diseases.
- Relate the life cycle of HIV in a helper T-lymphocyte to the processes of DNA replication, DNA transcription, protein synthesis and reverse transcription.
- Explain the link between standard of living and the incidence of disease.

Parasites and parasitism

Parasitism is a nutritional association between two organisms in which one — the **host** — is deprived of some nutrients, which the other — the **parasite** — gains. **Endoparasites**, such as *Plasmodium* and *Schistosoma*, live inside their host, while **ectoparasites**, such as fleas, live on their host.

Malaria

Key facts you must know

Malarial parasites are one-celled protoctistans belonging to the genus *Plasmodium*. Different species of *Plasmodium* cause different types of malaria. The most widespread is *Plasmodium falciparum*. Malaria is transferred to humans by female *Anopheles* mosquitoes (the **vector** of the parasite) when they bite people and suck their blood.

Tip Remember, the disease is caused by *Plasmodium*, not by the *Anopheles* mosquito.

The life cycle of the malarial parasite has two main phases:
- a sexual phase that occurs in the female *Anopheles* mosquito
- an asexual phase that occurs in human beings
- An infected female *Anopheles* mosquito bites a person and transfers the malarial parasite in its saliva before it sucks blood.
- The parasite travels in the bloodstream to the liver, where it infects liver cells. Here, for 6–10 days, it multiplies and changes form, but does not cause any symptoms.

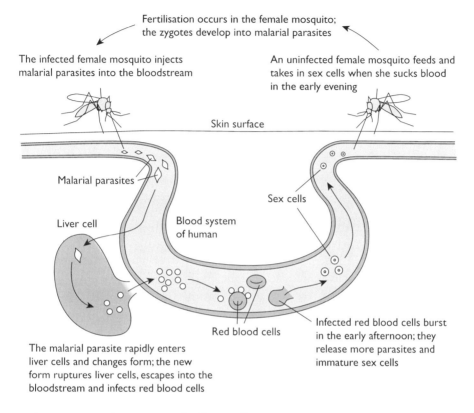

- The new parasites rupture the liver cells, enter the bloodstream and infect red blood cells.
- They multiply, producing more parasites and, sometimes, immature sex cells.
- The red blood cells burst, releasing the parasites and immature sex cells. This causes the fever that is typical of malaria. The cycle of infecting red blood cells and release of more parasites continues.
- The immature sex cells mature and are taken up by a female *Anopheles* mosquito when it sucks blood.
- Fertilisation takes place in the gut of the female mosquito and new malarial parasites are formed.
- The female mosquito bites another person and the cycle begins again.

Key concepts you must understand

The malarial parasite *(Plasmodium)* is adapted to its particular life cycle in the following ways:
- It has no locomotory structures.
- It is able to resist digestion in the gut of the female mosquito.
- It changes form in the liver, altering the antigens on its surface, which makes it difficult for the immune system to produce appropriate antibodies against the parasite.

- The parasite spends much of its life cycle 'hidden' inside either liver cells or red blood cells, which makes it difficult for the immune system to detect it.
- It reproduces rapidly in people and produces a large number of parasites.
- The immature sex cells are often released in the afternoon and mature 28 hours later — early evening — which is the most common feeding time for the female mosquito.
- The fever (produced by the bursting of the red blood cells) raises body temperature, which attracts the female mosquito.

Schistosomiasis

Key facts you must know

Schistosomiasis is sometimes called bilharzia. It is caused not by a microorganism, but by a **flatworm** (or **fluke**) called *Schistosoma*. Like *Plasmodium*, *Schistosoma* has two hosts in its life cycle — in this case, humans and some species of water snail.

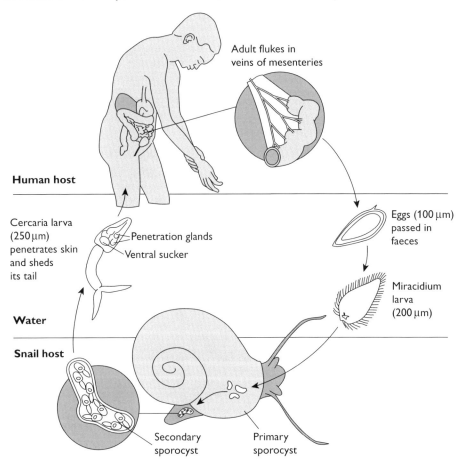

The main stages in the life cycle are as follows:
- Flukes in an infected person produce eggs that pass from the body in faeces and urine.
- If these eggs find their way into fresh water, they hatch into a larval stage called a **miracidium**.
- The miracidium infects a water snail.
- In the snail, the miracidium reproduces and changes into a second larval form called a **cercaria**.
- Cercariae escape from the snail into the water and penetrate the skin of people in the water.
- They enter the circulatory system and migrate to the veins of the bladder and intestines. In 6 weeks they mature into adult worms and begin producing eggs. (An adult worm may live for 25 years, producing more than 500 eggs per day — more than 4 000 000 eggs in total. Each egg could lead to many cercaria larvae being produced.)
- Some of the eggs then penetrate the walls of the veins, then those of the bladder or intestines. This causes considerable damage, resulting in the formation of much **scar tissue**. Eggs are then lost from the body in the urine or faeces and the cycle starts again.

However, many eggs do not penetrate the walls of the veins. They are carried in the bloodstream to the liver, where they are 'walled off' in a fibrous shell. The ongoing damage to intestines, bladder and liver accumulates over time and is often fatal. Annually, it is estimated that 2 000 000 people worldwide are infected by *Schistosoma*, resulting in 1 000 000 deaths.

Key concepts you must understand

Schistosoma is adapted to its particular life cycle in the following ways:
- There are no locomotory structures in the adult because its survival depends on remaining fixed to the walls of veins. It has no digestive system (products of human digestion are absorbed directly), circulatory system (its small size makes this unnecessary) or nervous system (nearly all responses are chemical). Therefore, most of the energy derived from nutrients is used in reproduction.
- The adult worm has suckers, which keep it in position in the veins (so that it is not washed away by the flow of blood).
- They avoid detection by the immune system by coating themselves with molecules derived from damaged red blood cells, thus appearing to be 'self' to the immune system.
- They produce vast numbers of offspring. This ensures survival of the species, because the life cycle depends entirely on 'chance' meetings between the larvae and the two hosts.
- A secondary host living in fresh water increases the chances of transmission because, in the countries where schistosomiasis is endemic, people spend much of their time near fresh water, bathing, washing clothes and watering animals.

AS Human Biology

What the examiners will expect you to be able to do
- Recall any of the key facts.
- Explain any of the key concepts.
- Use your knowledge and understanding of immune responses to explain in detail how *Plasmodium* and *Schistosoma* are difficult for the immune system to detect.

Defence against infection
General mechanisms
These non-specific mechanisms of defence do *not* involve the production of specific types of T-lymphocyte or B-lymphocyte in response to infection by a specific micro-organism.

Key concepts you must understand
There are two main types of general defence mechanism:
- excluding microorganisms from the body
- destroying microorganisms that have entered the body

Key facts you must know
Methods of excluding microorganisms are illustrated below.

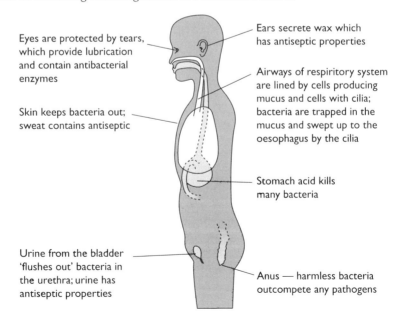

- Blood clotting allows damaged vessels and tissues to repair themselves while excluding microorganisms.
- Tears wash microorganisms from the surface of the eyes and contain antibacterial enzymes.
- Sweat is a mild antiseptic.
- Hydrochloric acid produced in the stomach kills many bacteria before they can enter the bloodstream.
- The action of cilia in the bronchioles sweeps bacteria out of the lungs.
- Commensal bacteria in the anus and rectum out-compete pathogenic bacteria for nutrients, preventing their multiplication.
- Urine has antiseptic properties and physically flushes bacteria out of the urethra.

Non-specific methods of destroying microorganisms include:
- fever — the high temperatures induced as a result of some infections are more harmful to the invading microorganisms than they are to our own body cells
- phagocytosis — several types of white blood cell engulf invading microorganisms and then destroy them

Tip Remember, a phagocyte is *not* a particular type of white blood cell; it is *any* cell capable of phagocytosis.

Blood clotting

Blood clotting is an example of an exclusion defence. It can occur as a result of:
- tissue damage — this leads to a pathway that results in scab formation and exclusion of microorganisms (the **extrinsic pathway**)
- damage to the endothelial lining of a blood vessel — this leads to clot formation inside a blood vessel (the **intrinsic pathway**) and is a major factor in heart disease

Extrinsic pathway of blood clotting

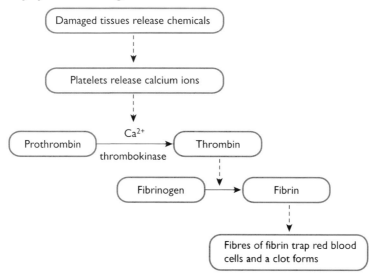

AS Human Biology

Blood clotting is initiated when platelets come into contact with damaged tissue as they leave the body (through the wound) with the blood. As they aggregate at the wound site, they become activated, releasing calcium ions. This causes a cascade of reactions, summarised in the diagram. Red blood cells become enmeshed in a 'net' of fibrin fibres and form a scab, which prevents both further blood loss and the entry of microorganisms. Underneath the scab, tissue repair takes place.

Phagocytosis

Phagocytosis is an example of destruction of microorganisms once they have entered the body. Several types of white blood cell are capable of phagocytosis. The most common are **neutrophils** and **macrophages**. When tissue damage takes place, chemicals called **mediators** are released, which attract phagocytes by **chemotaxis**. The capillary walls become more permeable and the phagocytes escape from the blood and move through the tissues.

They engulf bacteria by enclosing them in **phagosomes** formed by **pseudopodia** flowing around them. Once enclosed, the bacteria are destroyed by **lysosomes** releasing digestive enzymes and hydrogen peroxide into the phagosome.

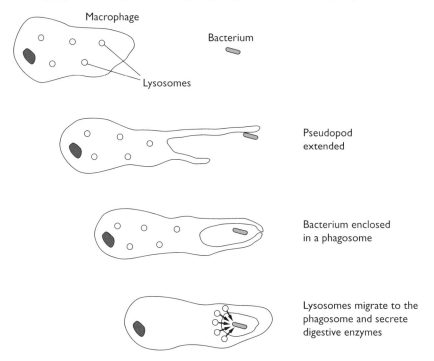

What the examiners will expect you to be able to do

- Recall any of the key facts.
- Explain any of the key concepts.
- Interpret diagrams and flowcharts representing blood clotting and phagocytosis.

Specific defence

Key concepts you must understand

The body makes specific **immunological** responses to specific organisms. **Immune responses** are brought about by T-lymphocytes and B-lymphocytes (T-cells and B-cells). T-lymphocytes attack pathogens directly, whereas B-lymphocytes produce **antibodies** to destroy them. Therefore, responses by T-lymphocytes are called **cell-mediated immunity** and responses by B-lymphocytes are called **antibody-mediated immunity** (or sometimes **humoral immunity**).

An immune response to a particular pathogenic microorganism is possible because the pathogen cells have specific markers on their surface called **antigens**. These are often glycoproteins and are different from the glycoproteins normally found in our bodies — they are **non-self** antigens.

Both B- and T-lymphocytes can be activated by these non-self antigens, but each B-lymphocyte or T-lymphocyte can only recognise one non-self antigen. This is because it has specific **antigen receptors** on its surface.

Antibodies are proteins (immunoglobulins) that can bind with antigens. This binding is very specific; each antibody recognises and binds with just one antigen.

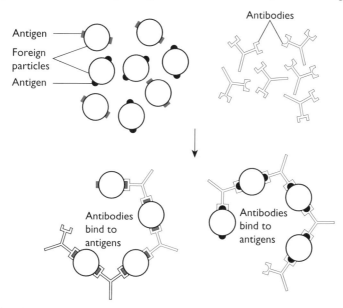

Immunisation
Immunisation means to make an individual immune to a disease.

Natural immunity arises as a result of natural processes:
- **Natural active immunity** results from immune responses to invading microorganisms.

- **Natural passive immunity** results when antibodies are *acquired*, not made as a result of infection. Antibodies are passed from mother to fetus across the placenta and from mother to baby in the colostrum and milk during breast-feeding.

Artificial immunity is acquired from processes that do not occur naturally:
- **Artificial active immunity** results from **vaccination**. During this process, the body is injected with an 'agent' that carries the same antigens as a particular pathogen. The body produces the same immune response as it would to that pathogen.
- **Artificial passive immunity** results from injecting specific antibodies directly.

Tip Be clear in your mind about the difference between the terms immunisation and vaccination. They are often used, incorrectly, to mean the same thing.

Key facts you must know

Antibody-mediated immunity

Each B-lymphocyte has the potential to produce a different antibody. When pathogenic microorganisms enter the body, some B-lymphocytes recognise the antigens on the surface of the microorganisms — receptors on their surface have a complementary shape to the antigens — and bind with them. These B-lymphocytes are activated and divide many times by mitosis to form a **clone** of millions of larger **plasma cells**. The plasma cells manufacture and release specific antibodies that bind with the antigens on the microorganisms. As they bind, they stick the microorganisms together, burst them open or destroy them in other ways. This is called the **primary immune response**.

Some of the original B-lymphocytes do not form plasma cells; they form **memory cells**. Memory cells do not take part in the destruction of microorganisms. They remain in the bloodstream and form the basis of **immunity** to that particular disease. If microorganisms of the same type enter the body again, the memory cells recognise their antigens, multiply and form plasma cells. This happens much more quickly than the initial response, because there are many more memory cells than there were original B-lymphocytes. The microorganisms are destroyed before they have had the time to become established and cause the disease. This is called the **secondary immune response**.

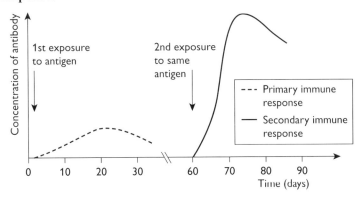

Cell-mediated immunity

Specific T-lymphocytes are activated in the same way as B-lymphocytes. Once activated, they divide many times by mitosis, but do not form plasma cells. As with B-lymphocytes, some T-lymphocytes become memory cells and take no part in the initial response. The remaining T-lymphocytes become **cytotoxic** and cause contact killing of target cells. These target cells could be cancer cells, cells infected by viruses or bacterial cells that have infected the body. The T-lymphocytes release chemicals that have two possible effects:
- They make holes in the cell membrane of the target cell, allowing the contents to leak out.
- They initiate a programmed cell-death process, the genetic code for which occurs in all cells.

Artificial active immunity

Vaccinated involves the injection of an 'agent' that carries the antigens of the microorganism (or its toxin) that causes the disease. This may be:
- an **attenuated** (weakened) strain of the microorganism (e.g. poliomyelitis, TB and measles vaccines)
- dead microorganisms (e.g. whooping cough and typhoid fever vaccines)
- modified toxins of the causal bacteria (e.g. tetanus and diphtheria vaccines)
- antigens alone (e.g. some influenza vaccines)
- harmless bacteria, genetically engineered to carry the antigens of pathogenic microorganisms (e.g. hepatitis B vaccine)

Using vaccines containing live microorganisms carries a slight risk. In the 1950s, a batch of polio vaccine contained virulent viruses because it had not been prepared properly. Children who were vaccinated with it contracted polio. Today, risks are better understood and procedures are more controlled. The risk is minimal.

Some microorganisms mutate frequently. As a result, the antigens on their surface change, making it virtually impossible to develop an effective vaccine. This is true of the virus causing the common cold. The influenza virus also mutates regularly, but not as quickly as that which causes the common cold. There is usually enough time to develop an effective vaccine before the virus mutates again.

What the examiners will expect you to be able to do
- Recall any of the key facts.
- Explain any of the key concepts.
- Interpret graphs of levels of antibody production in terms of the activity of B-lymphocytes.

Links You may need to use your knowledge of the tertiary structure of proteins from Module 1 to explain fully the specificity of the antibody–antigen binding process.

Non-communicable diseases

The biological basis of heart disease

Key concepts you must understand

A **heart attack** usually results from the sudden death of, or damage to, **cardiac muscle** in the wall of one of the ventricles of the heart — a **myocardial infarction**. Myocardial infarctions are usually the result of an interrupted blood supply to the cardiac muscle, which results in a lack of oxygen for aerobic respiration. Anaerobic respiration releases insufficient energy for the muscle cells, which die as a result. The most common cause of an interrupted blood supply is a blood clot or **thrombus** blocking one of the **coronary arteries**. The process of forming a blood clot is called **thrombosis**. If this happens in a coronary artery, it is a **coronary thrombosis**.

A blood clot need not necessarily remain in the artery in which it was formed. All, or part, of the clot may become dislodged and travel in the bloodstream to become lodged in another artery. It is then called an **embolus**. The blockage of another artery by this travelling clot is called **embolism**. The coronary arteries are narrower than most others, so emboli tend to become lodged there and cause myocardial infarctions.

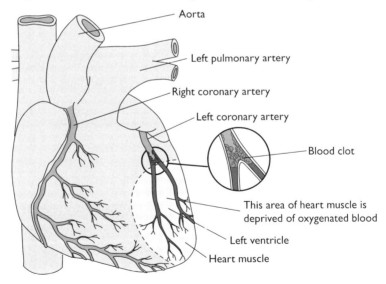

Key facts you must know

Atherosclerosis
Blood clots can form in any arteries as a result of atherosclerosis. In this process, a mixture of fatty substances, together called **atheroma**, becomes laid down under the endothelial lining of an artery.

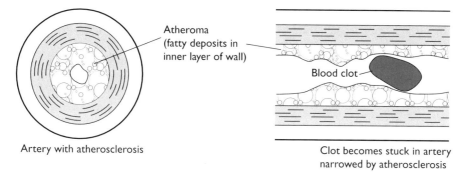

Artery with atherosclerosis

Clot becomes stuck in artery narrowed by atherosclerosis

The main substance in an atheroma is **cholesterol**. The atheroma subsequently **calcifies** (absorbs calcium) and becomes harder, forming an **atherosclerotic plaque**. The plaque affects the artery in three key ways:
- It narrows the artery, restricting blood flow — therefore reducing oxygen delivery — and increasing blood pressure.
- It makes it more likely that an embolus will become lodged in the artery.
- It promotes the intrinsic pathway of blood clotting as platelets brush against it.

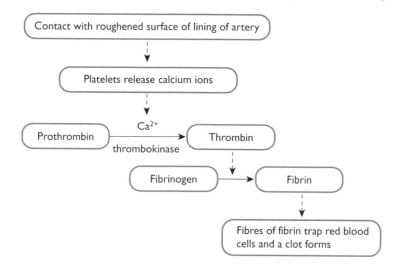

The plaque may also weaken the artery wall, forming an **aneurysm**.

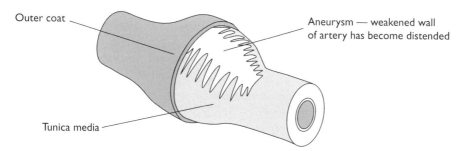

If the aneurysm bursts, serious complications can arise:
- Blood loss from a small aneurysm in an artery in the brain can put pressure on the brain and cause symptoms similar to those of a stroke.
- Blood loss from a large aneurysm in the aorta is nearly always fatal, as too much blood is lost too quickly.

A number of factors influence the incidence of coronary heart disease:

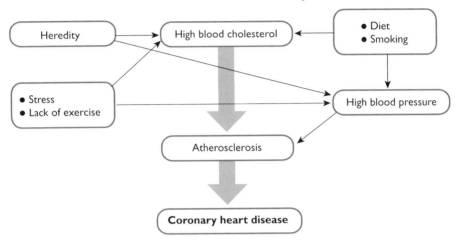

One of the most important factors is a genetic disposition towards high cholesterol levels and high blood pressure. You cannot change your genes! However, by modifying your lifestyle, you can reduce the extent to which other factors affect your overall risk.

The biological basis of cancer

Key concepts you must understand

Cancers are formed when a cell divides in an uncontrolled fashion. The cells derived from this form a **clone** of genetically identical cells that also divide in an uncontrolled way. Very soon, a mass of cells called a **tumour** is formed. Most tumours are **benign**. These are usually harmless, although they may cause problems because of *where* they grow. **Malignant** tumours divide in a more uncontrolled way and are much more dangerous. It is these tumours that we call **cancers**.

The genetic basis of cancer

Cell division is regulated so that it takes place at the rate required in a specific organ at a specific time. There are a number of control mechanisms that normally operate to prevent cell division from going out of control and forming tumours.

Proto-oncogenes are inactive genes that are present in all our cells. They can be transformed into active **oncogenes** in a number of ways. The gene may mutate, or it may be influenced by a viral infection. Various environmental factors are thought to increase the rate of mutation of oncogenes. These include **ionising radiation** and specific chemicals (cancer-makers or **carcinogens**), such as many of those found in tobacco smoke. Active oncogenes produce proteins that interfere with the normal regulation of cellular metabolism. The result is loss of control over cell division. In effect, a switch is set that says 'keep dividing'.

However, DNA that is damaged by mutation is often repaired by the action of proteins produced by other genes. Therefore, for a cancer to develop, this mechanism must fail too.

A third set of genes (**tumour-suppressor** genes) is also involved. These genes become active when a group of cells is dividing in an uncontrolled manner. They 'switch off' the division process. For a cancer to develop, these genes must fail as well.

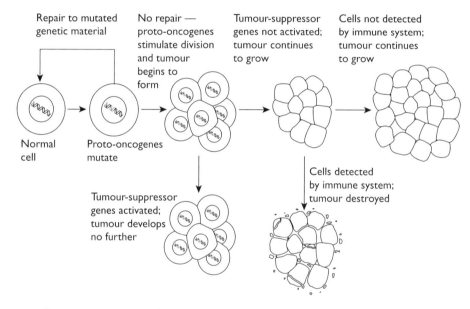

Key facts you must know

Benign tumours and malignant tumours differ in a number of important ways:
- Benign tumours usually grow much more slowly than malignant tumours.
- Benign tumours usually remain encased within a fibrous capsule and do not invade the tissue in which they originated. The boundaries of malignant tumours are much less defined and the cells frequently invade the tissue in which they originate.
- Benign tumours rarely show **metastasis**, i.e. spread to other parts of the body. Many malignant tumours do metastasise and cause **secondary cancers**.
- Malignant tumours often stimulate the development of a blood supply to the tumour.

For these reasons, it is often possible to remove benign tumours successfully by surgery, whereas complete removal of a malignant tumour is much more difficult and success much less certain. However, benign tumours can be dangerous. A benign tumour in the brain can exert pressure and affect brain function. For example, pressure

AS Human Biology

on the parts associated with the eyes can cause loss of vision. Pressure on certain motor areas can cause muscle weakness. Benign tumours may also exert pressure on the blood supply to an organ, restricting the blood flow.

The most common cancers in men and in women in the UK are shown in the table.

Men		Women	
Site	% of all cancers	Site	% of all cancers
Lung	21	Breast	19
Skin	14	Skin	11
Prostate	10	Lung	8
Bladder	6	Colon	6
Colon	6	Stomach	3
Stomach	5	Ovary	3
Rectum	5	Cervix	3
Lymph nodes	3	Rectum	3
Oesophagus	2	Uterus	2
Pancreas	2	Bladder	2
Other cancers	26	Other cancers	40

Tip Care must be taken in interpreting statistics such as these. First, look at what the statistics are actually showing — the *percentage of all cancers* in men and in women. There is no information about the *number* of cancers in men and in women. The data about lung cancer show that it represents 21% of all cancers in men and 8% of all cancers in women. However, if the total number of cancers in men were twice that in women, then the difference in the incidence of lung cancer would be greater than the table suggests. If there were more cancers overall in women than in men, the difference would be less than the table suggests. Second, are the differences linked to gender or do they reflect different lifestyles? Is skin cancer higher in men because they are male or because they take less care in protecting their skin from the effects of ultraviolet radiation?

What the examiners will expect you to be able to do

- Recall any of the key facts.
- Explain any of the key concepts.
- Interpret statistics concerning the incidence of heart disease and cancers.
- Relate the process of mitosis to the development of a tumour.

Links Many tumours form as a result of mutations to genes. These specific examples may be used as the basis of questions on mutation in Unit 5.

The cell cycle, mitosis and meiosis

The cell cycle

Key facts you must know

The cell cycle describes the sequence of events that occur as a cell grows, and prepares for and finally undergoes cell division by mitosis.

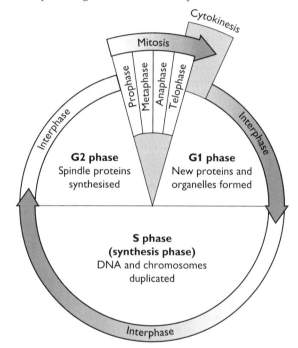

The phases of the cell cycle are as follows:
- **G1 phase** — the newly formed cell is comparatively small and starts to produce proteins and new organelles and begins to increase in size
- **S phase** — cell enlargement continues, the DNA of the cells replicates and each chromosome becomes a pair of **chromatids** joined by a **centromere**
- **G2 phase** — cell enlargement continues at a slower rate and the spindle proteins needed for cell division are synthesised
- **Mitosis** — the pairs of chromatids split into individual chromosomes and equal numbers of chromosomes move to opposite ends of the cell where each group forms a new nucleus
- **Cytokinesis** — the cell divides into two new cells

Some cells complete the cell cycle many times. They *must* divide repeatedly to form cells for new growth or for replacement of damaged cells. Examples include:
- cells in the bone marrow that divide repeatedly to produce red and white blood cells
- epithelial cells lining the intestine that divide repeatedly to replace those cells lost by being scraped off by food materials passing through the gut

Cancer cells also complete the cell cycle many times.

Cells that go through the cell cycle repeatedly are usually relatively unspecialised cells. Specialised cells often have a reduced capacity for cell division. For example, nerve cells, once formed, cannot divide.

Key concepts you must understand

The amount of DNA in the cell and the volume of the cell change throughout the cycle.

The DNA content doubles during the S phase as it replicates. It then returns to the normal level during cytokinesis as each newly formed cell receives half the chromosomes (and therefore half the DNA).

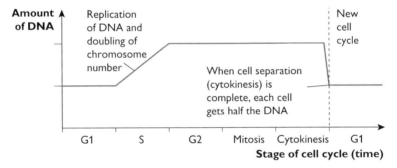

The volume increases steadily through G1, S and G2 as more and more proteins and cytoplasm are made. It then remains static during mitosis and then halves during cytokinesis as the cell divides into two.

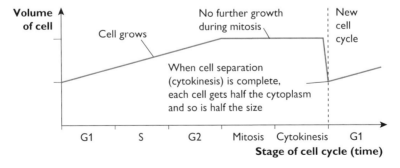

Specialised cells (like nerve cells), once formed, enter the G1 phase, produce the necessary proteins and organelles, grow to full size and then remain in this state without progressing to the S or G2 phase.

Mitosis

Key facts you must know

Mitosis results in two cells being formed with the same number and type of chromosomes as each other and as the parent cell which formed them. They are genetically identical — they form a clone of cells.

The process is divided into four key stages: **prophase**, **metaphase**, **anaphase** and **telophase**

Tip You do not need to know anything at all about the sub-divisions of prophase.

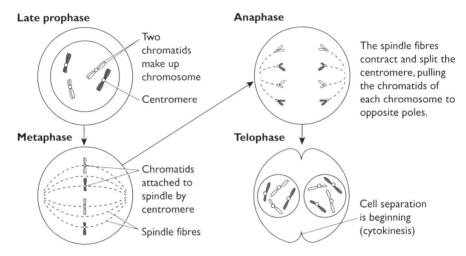

Stage of mitosis	Main events
Prophase	• Chromosomes coil and become visible as double structures — each is called a chromatid (during interphase, the DNA replicated and from one chromosome, two identical sister chromatids were formed) • Chromatids are held together by a centromere • The nuclear envelope starts to break down
Metaphase	• The spindle forms • The centromeres attach the chromatids to the spindle fibres so that they lie across the middle of the spindle
Anaphase	• The spindle fibres shorten and pull the sister chromatids to opposite poles of the cell • Once the chromatids have been separated, they become chromosomes again
Telophase	• The spindle fibres are broken down • The two sets of chromosomes group together at each pole and a nuclear envelope forms around each • The chromosomes uncoil and cannot be seen as individual structures

Key concepts you must understand

Strictly, mitosis describes the division of the chromosomes, *not* of the whole cell. The division of the cell is called cytokinesis.

The 46 chromosomes in nearly all human cells are, in fact, 23 pairs called **homologous pairs**. One chromosome from each pair is paternal in origin (from the father) and the other is maternal in origin (from the mother).

Cells with pairs of all the homologous chromosomes like this are called **diploid** cells. The diploid condition is sometimes written as **2n** — 'n' is the number of different chromosomes and there are two (a homologous pair) of each.

Homologous chromosomes carry genes for the same features in the same sequence, although they may not carry the same **alleles** (versions) of these genes.

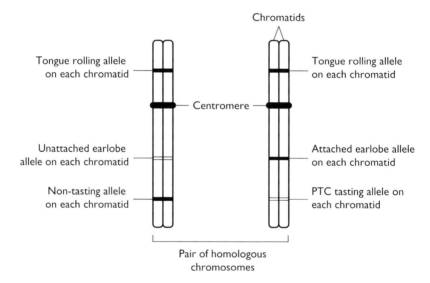

Homologous chromosomes are the same size and shape because they carry the same genes in the same sequence. The sister chromatids, formed during interphase, *always carry the same alleles of the genes* because of the semi-conservative replication of DNA in the original chromosome.

The cells formed by mitosis are identical because each receives one chromatid from each pair of sister chromatids.

Tip So what *is* all this about chromosomes and chromatids? A chromosome usually contains just *one* molecule of DNA and some protein molecules (mainly histones) bound to it. The diagrams below show what happens to a chromosome during the S phase of interphase. Each of the two structures is like a chromosome since it contains DNA and proteins. However, chromosomes don't normally join together, so we call these structures with one DNA molecule **chromatids**; the whole double structure is still a chromosome. Once they have separated during mitosis, the chromatids are chromosomes again!

AQA(A) Unit 3

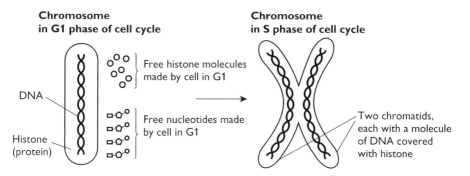

Following fertilisation, the zygote formed is 'copied' repeatedly by mitosis so that all the body cells are diploid and genetically identical. This underlines the importance of mitosis in life cycles.

Meiosis

Key facts you must know

- Meiosis produces cells that show genetic variation.
- Meiosis halves the chromosome number of cells (in human beings from 46 to 23).
- The cells formed by meiosis are usually sex cells.
- There are two cell divisions involved in meiosis.
- Meiosis produces cells that contain only one chromosome from each homologous pair. Such cells are called **haploid** cells. The haploid condition is sometimes written as 'n'.

Key concepts you must understand

Production of haploid gametes (sex cells) is a necessary part of a life cycle so that when fertilisation occurs, the normal diploid number is restored.

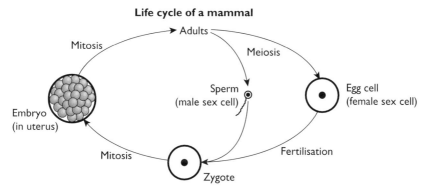

Tip In an exam question, diagrams like the one above often include chromosome numbers. Meiosis can be identified by the halving of the chromosome number.

What the examiners will expect you to be able to do
- Recall any of the key facts.
- Explain any of the key concepts.
- Be able to sequence drawings showing cells in various stages of mitosis.
- Compare and contrast meiosis and mitosis.
- Interpret graphs giving information about features of the cell cycle such as cell volume, DNA content and distance of chromosomes from the poles of a cell.
- Recognise from drawings or photographs the various stages of the cell cycle.
- Interpret diagrams of unfamiliar life cycles to identify stages where meiosis and mitosis are taking place.
- From information given about unfamiliar life cycles, deduce the number of chromosomes per cell at various stages of the cycle.

Links An understanding of meiosis is crucial to the understanding of the mechanisms of inheritance. Meiosis in sexual reproduction is a source of genetic variation, which is essential to the process of evolution.

The structure and functions of DNA and RNA

The structure of DNA and RNA

Key facts you must know
- DNA and RNA are **nucleic acids**, built from **nucleotides**. Because DNA and RNA contain many nucleotides, they are called **polynucleotides**.
- Each nucleotide consists of a **nitrogenous base**, a **pentose sugar** and a **phosphate** group.

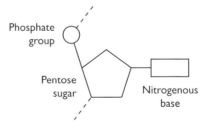

DNA
- DNA is a double-stranded molecule.
- Each strand of DNA is built from four types of nucleotide. Each nucleotide differs only in the base it contains. The bases in DNA nucleotides are **adenine**, **thymine**,

cytosine and guanine, abbreviated to **A**, **T**, **C** and **G**. The pentose sugar in DNA nucleotides is **deoxyribose**.
- Adenine on either strand is always opposite thymine on the other. Cytosine on either strand is always opposite guanine on the other. This specific pairing is called the **base-pairing rule**. Adenine is said to be **complementary** to thymine; cytosine is complementary to guanine.
- Because of the base-pairing rule, there are equal amounts of adenine and thymine and equal amounts of cytosine and guanine in *every* DNA molecule.
- **Hydrogen bonds** between complementary bases hold the two strands together.

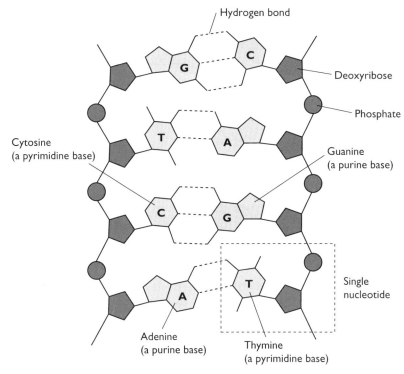

Note: adenine (A) always bonds with thymine (T)
guanine (G) always bonds with cytosine (C)

- The two strands of the DNA molecule are coiled into a helix. This double helical structure was first proposed by James Watson and Francis Crick in 1953.
- DNA is found in chromosomes in the nuclei of eukaryotic cells. In prokaryotic cells, there is no distinct nucleus; circular DNA is free in the cytoplasm.
- Each chromosome contains one molecule of DNA, bound to proteins.
- DNA specifies the code for protein synthesis.

RNA
- There are three types of RNA:
  - **messenger RNA (mRNA)** carries the DNA code for protein synthesis to the ribosomes

AS Human Biology

- **transfer RNA** (**tRNA**) transfers free amino acids to the ribosomes
- **ribosomal RNA** (**rRNA**) is a structural component of the ribosomes
• mRNA is a single-stranded linear molecule.
• tRNA is a single-stranded molecule conformed into a cloverleaf shape.

• All three types of RNA differ from DNA in several ways:
  – they are all single-stranded molecules
  – they are all much smaller molecules than DNA
  – the base thymine is replaced by the base **uracil**
  – the pentose sugar is **ribose** (not deoxyribose)

Tip Although tRNA is a single-stranded molecule, there *is* still some base pairing. In regions of the molecule where the strand folds back on itself, hydrogen bonds between complementary bases hold the molecule in shape. Beware of writing that RNA never shows any base-pairing.

DNA replication

Key facts you must know

• DNA replication occurs during the S phase (interphase) of the cell cycle.
• DNA replication is semi-conservative.

```
 ┌T A┐
 ├A T┤
 ├T A┤
 ├C G┤
 ├G C┤
 └A T┘
```
DNA helicase separates the polynucleotide strands of DNA

```
┌T A┐  ┌T A┐
├A T┤  ├A T┤
├T A┤  ├T A┤
├C G┤  ├C G┤
├G C┤  ├G C┤
└A T┘  └A T┘
```
...each strand acts as a template for the formation of a new molecule of DNA...

...DNA polymerase assembles DNA nucleotides into two new strands according to the base-pairing rule...

```
┌T A┐  ┌T A┐
├A T┤  ├A T┤
├T A┤  ├T A┤
├C G┤  ├C G┤
├G C┤  ├G C┤
└A T┘  └A T┘
```
...two identical DNA molecules are formed — each contains a strand from the parent DNA and a new complementary strand

Key concepts you must understand

- Replication of DNA is called semi-conservative because one of the strands of the original DNA molecule is conserved in each new molecule of DNA.
- The two molecules of DNA formed are identical to each other, and to the original molecule. This is because of the base-pairing rule, which ensures that each new strand formed has the same base sequence as the strand 'lost' from the molecule when the two strands split apart.

Protein synthesis

Key facts you must know

- Proteins are built from chains of amino acids linked by peptide bonds (see Unit 1). This takes place in the **ribosomes**.
- The sequence of amino acids in a protein molecule is coded for by the sequence of bases in a gene in a DNA molecule.
- mRNA carries this code from the DNA to the ribosomes.
- tRNA carries the free amino acids to the ribosomes, where they are assembled into proteins.

The diagram below gives an overview of the stages of protein synthesis.

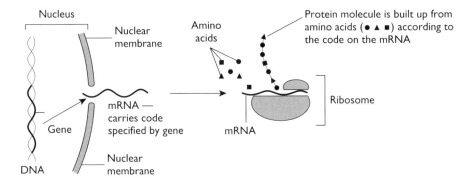

Key concepts you must understand

The DNA code
- The DNA code is a **triplet** code — a sequence of three bases codes for one amino acid.
- The DNA code is **universal** — the same triplets of bases code for the same amino acids in all organisms.
- The DNA code is **degenerate**. There are 64 possible combinations of the three bases and only 20 amino acids, so some amino acids have more than one code and some triplets do not code for amino acids but carry other information, such as a signal to 'stop' because the sequence ends here.

First position	Second position				Third position
	T	C	A	G	
T	PHE	SER	TYR	CYS	T
	PHE	SER	TYR	CYS	C
	PHE	SER	stop	stop	A
	PHE	SER	stop	TRP	G
C	LEU	PRO	HIS	ARG	T
	LEU	PRO	HIS	ARG	C
	LEU	PRO	GLN	ARG	A
	LEU	PRO	GLN	ARG	G
A	ILE	THR	ASN	SER	T
	ILE	THR	ASN	SER	C
	ILE	THR	LYS	ARG	A
	MET	THR	LYS	ARG	G
G	VAL	ALA	ASP	GLY	T
	VAL	ALA	ASP	GLY	C
	VAL	ALA	GLU	GLY	A
	VAL	ALA	GLU	GLY	G

- Some DNA is **non-coding** — there are regions that do not code for anything.

AQA (A) Unit 3

Transcription

The DNA triplet code can be **transcribed** (written in another form) into a triplet mRNA code. Each mRNA triplet is called a **codon** and the bases it contains are complementary to those in a DNA triplet. The main stages in transcription are as follows:

- The section of DNA that is the gene coding for the particular protein unwinds.
- RNA polymerase assembles an mRNA strand from free RNA nucleotides according to the base-pairing rule, but uracil is used in the RNA molecule instead of thymine.
- Any non-coding regions of the mRNA molecule (transcribed from non-coding DNA) are now 'cut out' by enzymes.
- The 'finished' mRNA molecule leaves the nucleus through a pore in the nuclear envelope.

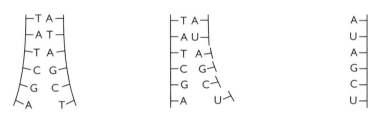

DNA molecule in nucleus unzips

RNA polymerase joins RNA nucleotides together to form mRNA molecule

mRNA molecule leaves nucleus and becomes attached to ribosomes

Translation

Translation occurs in the ribosomes. The mRNA molecule feeds through the ribosome and its code is 'translated' into a chain of amino acids.

As the mRNA molecule feeds through the ribosome, the following events take place:

- The first two codons on the mRNA molecule position themselves inside the ribosome.
- tRNA molecules with complementary **anticodons** bind with the first two mRNA codons, and so carry their amino acids into position.
- A peptide bond forms between these two amino acids.
- The mRNA moves along by one codon.
- The third mRNA codon is now in the ribosome, and tRNA with a complementary anticodon binds to it, bringing its amino acid into position. The first tRNA has been moved out of the ribosome and breaks away.
- A peptide bond forms between the second and third amino acids.
- The mRNA moves along by one codon.
- The fourth mRNA codon is now in the ribosome, the second tRNA has been moved out and breaks away...and so on until the last codon (**stop codon**) is in position and translation ceases.

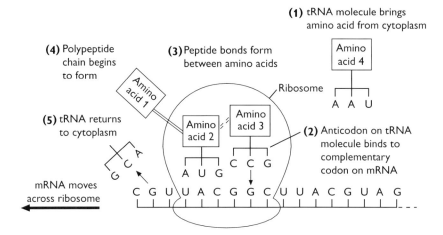

DNA codes for all the enzymes that are synthesised. Therefore, it controls indirectly all the reactions occurring in cells. Some genes are only active in certain cells and so the proteins coded for by those genes are only manufactured in those cells.

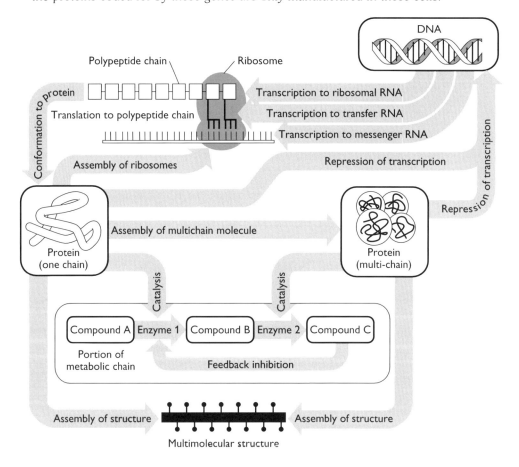

AQA (A) Unit 3

What the examiners will expect you to be able to do
- Recall any of the key facts.
- Explain any of the key concepts.
- Identify components of a DNA molecule from a drawing.
- Identify missing bases in a drawing of DNA (use the base pairing rule — A pairs with T and C pairs with G).
- Calculate percentages of each base in a DNA molecule given the percentage of just one base.

Tip Remember, because of the base-pairing rule, A = T and C = G *and* A + T + C + G = 100%. So, if A = 26%,
T is also 26%
A + T = 52%
C + G = 100% − 52% = 48%
C = G, so the value of each is 48/2 = 24%

- Describe the sequence of an mRNA strand formed by transcription of a given DNA sequence. (Remember that U is used instead of T in RNA and apply the base-pairing rule.)

Links You will probably need to apply your understanding of DNA structure to areas such as meiosis, genetic variation, mutations and natural selection. In plants, protein synthesis is related to photosynthesis, synthesis of amino acids and mineral nutrition.

Genetic engineering
Using gene technology to combat disease

This section looks at the ways in which the genotype of an organism can be modified by the introduction of genes from a different organism and how scientists can be sure that the appropriate genes have been transferred. It also considers whether such practices are ethical.

Key facts you must know
- A section of DNA that codes for a particular polypeptide (or protein) is a **gene**.
- The total complement of genes in an organism is called the **genome**.
- Biotechnologists can show the presence of a gene in a sample of DNA by using a gene-specific **DNA probe** (**gene probe**).
- Genes can be 'cut out' of a DNA molecule using **restriction endonuclease** enzymes.
- Genes can be inserted ('tied') into another DNA molecule using **ligase** enzymes.

Tip To help you get the names the right way round, remember that *ligation* means tying — think of the *ligature* used in tying back together a surgical wound.

- Using an enzyme called **reverse transcriptase**, genes can be made starting from a molecule of RNA. They can then be inserted into a molecule of DNA.
- DNA that has had new genes added to it is called **recombinant DNA**.
- Genes are transferred into other cells using **vectors**. These are usually either **plasmids** (small, circular pieces of DNA found in bacteria) or viruses.

Tip A vector is an agent of transfer. The female *Anopheles* mosquito is the vector for the malarial parasite and houseflies are vectors of disease-causing microorganisms. The gene must be inserted into the vector and then the vector transfers the gene to the desired cell.

Biotechnologists need to check whether microorganisms have actually taken up plasmids. They do this by seeing if they are resistant to a certain antibiotic. The resistance gene is transferred into the microorganism on the same plasmid, so any microorganisms that are resistant must have taken up the plasmid (and so must have the desired gene too).

An organism that has had DNA from a different organism added to it is called a **transgenic** organism. By creating appropriate recombinant DNA and inserting it into a microorganism, a transgenic organism can be created that will produce a useful substance, such as human growth hormone or human insulin.

Many people are concerned about the new gene technology and see potential problems such as:
- When a biotechnology company 'creates' a new gene, whose gene is it?
- The technology may accidentally create harmful genes or transfer harmful genes into bacteria used to produce a product for human beings. The harmful product of these genes may find its way into humans.
- Biotechnologists are tampering with God's creation.
- Genetically modified crop plants are 'not natural'. The genes from these plants might be transferred into other, related, wild plants.

Key concepts you must understand

When biotechnologists produce transgenic microorganisms to manufacture a certain product, they must first identify the gene (DNA sequence that codes for the desired product) in the 'donor' cells and remove it from these cells. Alternatively, they could create the gene from an appropriate mRNA molecule using reverse transcriptase. They must then transfer the gene to the microorganism using a vector and culture these transgenic microorganisms to give large numbers, so that significant amounts of the product can be synthesised.

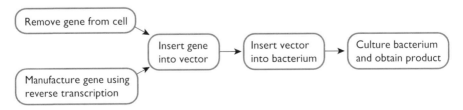

Identifying the gene

To check that a particular gene is present in the donor cells, biotechnologists create a DNA probe (sometimes also called a gene probe). This is a section of single-stranded DNA with a base sequence that is complementary to that of the gene. The probe is usually made radioactive or fluorescent so that it can be detected.
- The DNA in the donor cells is extracted, split into single strands and mixed with the DNA probe.
- The DNA probe binds with those sections of DNA that have a complementary base sequence to that of the probe.
- Excess probe is washed off and the remaining DNA is checked for radioactivity (or fluorescence). If it is radioactive (or fluorescent), this means that the probe has bound to some complementary DNA. The gene must be present.

Removing the gene from the donor cell

Donor cells are incubated with restriction endonucleases. These enzymes cut the DNA at specific base sequences. By selecting the restriction enzymes carefully, a section of DNA containing the required gene can be isolated.

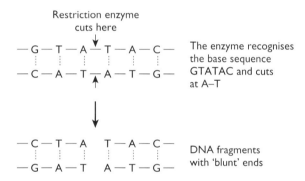

Some restriction enzymes do not make a 'clean cut' across the two strands of the DNA, but make a staggered cut leaving unpaired bases. These staggered ends are often called '**sticky ends**'.

Creating the gene from mRNA

If the mRNA molecule that codes for the polypeptide (protein) is available, the gene can be synthesised using reverse transcriptase in the following way:

- The mRNA molecule coding for the protein is incubated with reverse transcriptase and the necessary free nucleotides.
- Reverse transcriptase creates a single strand of complementary DNA (cDNA).
- The mRNA is 'washed' out.
- The DNA is incubated with DNA polymerase and free DNA nucleotides.
- DNA polymerase creates a complementary strand of DNA which bonds with the strand created by reverse transcriptase. (Check the section on DNA replication for further detail). This double-stranded DNA version of the gene can now be transferred to another cell.

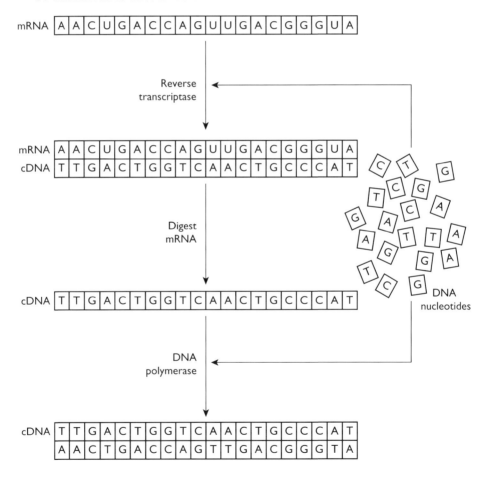

Tip Reverse transcriptase is a very logical name for the enzyme. Think of what happens during transcription — a section of one strand of DNA is 'transcribed' to make an mRNA molecule. Throw it into reverse (reverse transcription) — a short, single strand of DNA is made from mRNA.

AQA (A) Unit 3

Transferring the gene

Genes are transferred using plasmids. The diagram below shows the main stages in the process.

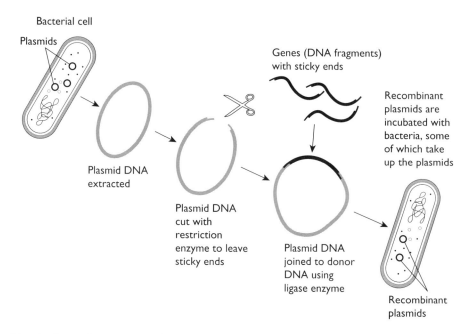

Tip It all sounds very straightforward — snip the gene out with molecular scissors or create the gene with a biochemistry set, glue it into a plasmid and get the bacteria to 'swallow up' the plasmids. However, you must realise that there is a very small percentage success rate. For example, when the bacteria are incubated with the plasmids, as few as 1 bacterium in 100 000 may actually take up the plasmid! So how do we know that *any* of them have?

Checking that the bacteria have taken up the plasmid

There are a number of ways that this can be done, but the most usual method relies on antibiotic resistance. The procedure is as follows:
- Transfer to the plasmids not only the desired gene, but also a gene that confers resistance to a particular antibiotic.
- Incubate the bacteria with the plasmids.
- Culture the bacteria on a medium containing the antibiotic.

Only those bacteria that are resistant to the antibiotic can survive. To be resistant, they must have taken up the plasmid, so they must *also* have the desired gene.

Culturing the transgenic microorganisms

Once the transgenic bacteria have been identified, they can be transferred from the antibiotic-containing medium and cultured in a liquid medium, initially on a relatively small scale. As the bacteria reproduce, so does the plasmid containing the new gene. The new gene is copied into all the bacteria — we say that the gene has been **cloned**. Once the bacteria have multiplied, an **inoculum** from this culture is transferred to a much larger fermenter where the bacteria multiply and produce the desired substance (e.g. human insulin or human growth hormone). The product is isolated and purified by downstream processing.

What the examiners will expect you to be able to do

- Recall any of the key facts.
- Explain any of the key concepts.
- Complete flow charts showing the different stages of genetic manipulation.
- Interpret information given about an unfamiliar example of genetic engineering.
- Comment, from data given, on possible advantages and disadvantages of a particular process.

Links Advances in genetic engineering are being made daily and genes are being transferred into many organisms to try to combat genetic disorders. Cystic fibrosis is just one example of **gene therapy** — genes are introduced into human cells to try to combat the disorder. Cells from the pituitary gland in mice have been genetically engineered to produce insulin and reintroduced into diabetic mice to control their diabetes. Plants are being genetically engineered to make them more able to survive in particular environments. You might be asked about genetic engineering in other modules, where such topics are covered.

Diagnosis and treatment of disease

Using DNA probes for diagnosis

Key facts you must know

DNA probes are used in the diagnosis of genetically determined diseases. These include cystic fibrosis, Huntington's disease and some forms of haemophilia. It is possible to use gene probes to screen individuals for the presence of such genes. The main stages in the process are as follows:

- DNA is extracted from white blood cells.
- It is cut into smaller sections (fragments) using restriction enzymes.
- The DNA fragments are separated by **gel electrophoresis**.
- They are then transferred to a nylon membrane.
- Radioactive DNA probes are used to locate specific DNA base sequences and thus show the positions of the DNA fragments.
- An X-radiograph is prepared of the separated DNA fragments.
- The pattern of bands produced on the X-radiograph is the **genetic fingerprint** of that person.

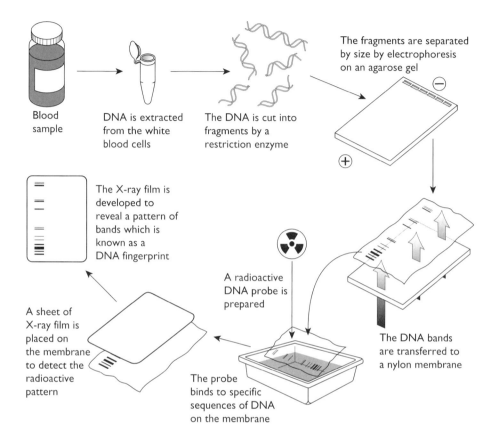

Key concepts you must understand

Gel electrophoresis separates fragments of DNA on the basis of the charge on the molecule and the size of the fragment. It is the 'sieving' effect of the gel that limits how easily molecules can move through it. Therefore, the *size* of the DNA fragment largely determines how far it moves. The molecular masses of the fragments depend on their size and can be calculated from the distance moved.

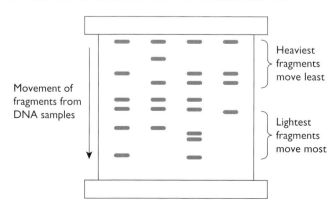

Once the separation is complete, the fragments are carefully transferred to a membrane before having DNA probes applied. After washing the probe out, the regions where the probes have bound are displayed using X-ray film, which becomes 'fogged' when exposed to radioactivity. The resulting pattern is the DNA fingerprint.

By comparing the fingerprints of several people known to suffer from a genetically determined condition with those of people known *not* to suffer from the condition, the DNA sequences associated with that disease can be deduced.

Using enzymes for diagnosis

Key concepts you must understand

Enzymes can be used to diagnose specific non-communicable diseases. Knowledge of these enzymes is useful in two ways:
- Some diseases (e.g. **pancreatitis**) cause changes in the concentration and distribution of some of the enzymes found naturally in our bodies.
- Enzymes are specific, and can therefore be used as analytical reagents to show the presence of specific compounds associated with disease (e.g. altered glucose levels in **diabetes**).

Key facts you must know

Pancreatitis

Pancreatitis (inflammation of the pancreas) results from a blockage of the pancreatic duct. As a consequence, pancreatic enzymes cannot escape into the duodenum. These enzymes 'back up' the pancreatic duct where they digest pancreatic tissue and escape into the bloodstream. The abnormal presence of pancreatic enzymes in the blood plasma is used to diagnose the condition.

Acute (short-term) pancreatitis causes digestive problems, because pancreatic amylase, proteases and lipase are unable to act on the food entering the duodenum. Chronic (ongoing) pancreatitis can, in addition, result in permanent damage to the pancreas as fibrous scar tissue is formed to replace damaged pancreatic tissue. Pancreatitis can result from:
- alcohol abuse
- some viral infections (e.g. mumps)
- cystic fibrosis

Diabetes

The plasma glucose level in a diabetic can rise above the **renal threshold**; as a result, glucose appears in the urine. Enzyme-based analysis of the glucose in the urine allows a semi-quantitative estimation of the level of plasma glucose.
- The enzymes **glucose oxidase** and **peroxidase** are immobilised on a paper strip.
- Glucose oxidase oxidises glucose to gluconic acid and hydrogen peroxide.

- In the presence of peroxidase, the hydrogen peroxide oxidises a colourless dye embedded in the paper strip.
- The resulting colour represents concentration range.

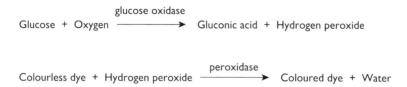

Using antibiotics to treat bacterial diseases

Key concepts you must understand

Antibiotics act against bacteria by disrupting cellular processes such as:
- DNA replication
- protein synthesis
- cell wall synthesis

Some antibiotics kill bacteria — these are **bactericidal** antibiotics. Others do not kill bacteria but stop them from reproducing — these are **bacteriostatic** antibiotics.

Key facts you must know

The ways in which some different antibiotics act are summarised in the table below.

Mode of action of antibiotic	Example	How the antibiotic works	Bactericidal or bacteriostatic?
Disrupts cell wall synthesis	Penicillin	Weakened cell wall cannot resist entry of water by osmosis and cell bursts (osmotic lysis)	Bactericidal
Disrupts DNA replication	Nalidixic acid	Bacteria are not killed, but cell division is halted	Bacteriostatic
Disrupts protein synthesis	Tetracycline	Bacterial cell cannot synthesise enzymes and structural proteins	Bactericidal

Antibiotics that disrupt cell wall synthesis interfere with the synthesis of the **peptidoglycan** layer in the cell wall. Water enters by osmosis down a water potential gradient. Ordinarily, this entry would be resisted by the cell wall. With only an incomplete cell wall to resist the swelling caused by the entry of the water, the bacterial cell bursts.

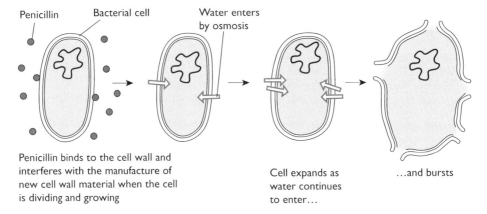

Penicillin binds to the cell wall and interferes with the manufacture of new cell wall material when the cell is dividing and growing

Cell expands as water continues to enter…

…and bursts

Antibiotics are used almost entirely to treat diseases caused by bacteria. Viruses are acellular and so do not carry out metabolic processes while outside human cells. Once inside human cells, they are protected by that cell — antibiotics cannot enter, as there are no transport proteins to carry them in. Therefore, antibiotics cannot be used to treat viral diseases. They are sometimes prescribed for viral conditions in order to prevent a secondary bacterial infection.

Using beta-blockers to treat hypertension

Key facts you must know

Hypertension, or sustained high blood pressure, is a major risk factor in coronary heart disease. It promotes narrowing of arteries by atherosclerosis. Although underlying causes are often difficult to pinpoint, an immediate cause of hypertension is sustained secretion of **adrenaline** and **noradrenaline** at the endings of neurones in the **sympathetic nervous system**. These two **neurotransmitters** bind to specific **beta-receptors** found in the **cardiac muscle** of the ventricles, as well as in the walls of arteries and arterioles and airways in the lungs. Their secretion is part of the normal **stress response** and causes:
- the ventricles to contract faster and with more force
- the airways in the lungs to dilate
- the arteries and arterioles leading to cardiac and skeletal muscle to dilate, while many others constrict

The overall effect of this is to allow more oxygenated blood to be circulated around the body. This is desirable in times of physical stress, such as exercise. However, it is undesirable when there is no need for physical exertion. Beta-blocker molecules have structures that will also bind with beta-receptors. They therefore reduce the effects of adrenaline and noradrenaline by competing for the beta-receptor sites.

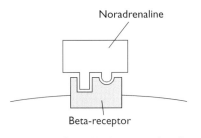
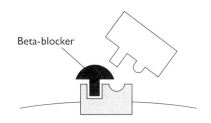

In the absence of beta-blockers, noradrenaline can bind to beta-receptors to increase heart rate and force of contraction

In the presence of beta-blockers, noradrenaline is prevented from binding to beta-receptors

Some beta-blockers are **non-selective** and bind to all beta-receptors. They prevent all the effects described. Others are **cardio-selective** and bind only to the beta-receptors in the cardiac muscle in the ventricles. These reduce the rate and force of contractions (and so reduce blood pressure) but do not affect the airways, arteries and arterioles. This means that oxygen intake can still be increased during exercise.

Monoclonal antibodies

Key concepts you must understand

When microorganisms enter our body, they stimulate an immune response (see pp. 29–31). There are several different marker antigens on the surface of each microorganism and each can activate a different type of B-lymphocyte. Each type of B-lymphocyte divides many times to form a clone of plasma cells, which produce antibodies against one of the antigens. Because clones of several different plasma cells are involved in producing antibodies against one microorganism, these antibodies are called **polyclonal antibodies**.

Monoclonal antibodies are the product of a single type of plasma cell derived from a single type of B-lymphocyte. Monoclonal antibodies bind with one specific antigen only. They can distinguish between antigens that differ only slightly.

Key facts you must know

Monoclonal antibodies, because of the precision of their binding, can be used in a number of ways.

Detection of some types of cancer cell

B-cells are activated using the cancer-cell antigens. They then produce monoclonal antibodies. Other potentially cancerous cell samples can then be tested with the monoclonal antibodies. If the antibodies bind to the cells, this shows that the cells must have the cancer-cell antigens and must therefore be cancerous. Binding can be detected by making the antibodies fluorescent, in the same way that gene probes are made fluorescent.

Detection of some viruses

This works in essentially the same way as detecting cancer cells. It is an important tool in screening blood samples for viruses such as HIV.

Pregnancy testing

As soon as an embryo implants in the uterus, it begins to produce a hormone called **hCG (human chorionic gonadotrophin)**. hCG is only produced under these circumstances, so it is a certain indicator of pregnancy. Some hCG is present in a pregnant woman's urine and so a sample can be tested for its presence. The diagram shows how monoclonal antibodies, produced in response to exposing lymphocytes to hCG, are used in a diagnostic pregnancy test.

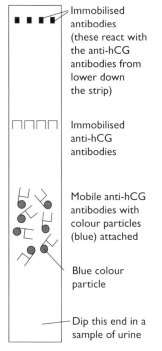

Treatment of cancers

Research is currently underway into binding anti-cancer drugs to monoclonal antibodies. At the moment, conventional chemotherapy often targets any actively dividing cells. This includes the cancer cells, but these drugs also target:
- cells in the bone marrow that are dividing to produce red and white blood cells
- cells lining many hollow organs, such as the intestines and stomach
- cells in hair follicles

Hopefully, manufacturing monoclonal antibodies in response to the antigens on the cancer cells and then binding the anti-cancer drug to the antibodies will mean that the drug can be delivered to the cancer cells only. This will mean that much higher drug doses can be used without side effects.

What the examiners will expect you to be able to do

- Recall any of the key facts.
- Explain any of the key concepts.
- Interpret unfamiliar data concerning the action of antibiotics.
- Interpret data showing the effects of different types of beta-blocker on blood pressure.
- Interpret flow charts showing the production of monoclonal antibodies.

This section contains questions similar in style to those you can expect to see in your Unit 3 examination. The limited number of questions means that it is impossible to cover all the topics and all the question styles, but they should give you a flavour of what to expect. The responses that are shown are real students' answers to the questions.

There are several ways of using this section. You could:
- 'hide' the answers to each question and try the question yourself. It needn't be a memory test — use your notes to see if you can actually make all the points you ought to make
- check your answers against the candidates' responses and make an estimate of the likely standard of your response to each question
- check your answers against the examiner's comments to see if you can appreciate where you might have lost marks
- check your answers against the terms used in the question — did you *explain* when you were asked to, or did you merely *describe*?

Examiner's comments

All candidate responses are followed by examiner's comments. These are preceded by the icon 🖉 and indicate where credit is due. In the weaker answers, they also point out areas for improvement, specific problems and common errors such as lack of clarity, weak or non-existent development, irrelevance, misinterpretation of the question and mistaken meanings of terms.

Question 1

The structure of DNA and RNA

(a) Figure 1 represents the structure of the DNA molecule.

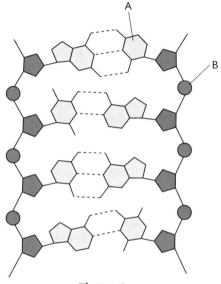

Figure 1

(i) Name the structures labelled **A** and **B**. (2 marks)

(ii) Use the diagram to explain why the DNA molecule is sometimes described as consisting of two polynucleotide strands. (1 mark)

(b) Complete the table to show three differences in structure between **DNA** and **mRNA**.

DNA	mRNA

(3 marks)

Total: 6 marks

■ ■ ■

Candidates' answers to Question 1

Candidate A
(a) (i) A — nucleotide; B — phosphorus

Candidate B
(a) (i) A is a nitrogenous base (adenine, thymine, cytosine or guanine);
B is a phosphate group which links adjacent nucleotides.

question 1

e Candidate A is clearly confused between nucleotide and nitrogenous base — make sure you aren't, it's not a difficult question. Phosphorus is a chemical element and is not an acceptable answer. B is a phosphate group. Candidate B is clearly familiar with the components of a DNA molecule and is awarded both marks.

Candidate A

(a) (ii) There are two strands with lots of nucleotides joined.

Candidate B

(a) (ii) Each strand consists of many nucleotides linked by the phosphate groups. 'Poly-' means many, like in a polygon.

e Both candidates clearly understand the idea, for 1 mark.

Candidate A

(b)

DNA	mRNA
is made of two strands	only has one strand
has uracil instead of thymine	has thymine instead of uracil
has a different sugar	has a different sugar

Candidate B

(b)

DNA	mRNA
DNA is a double helix	RNA is just a single helix
The pentose sugar is deoxyribose	The pentose sugar is ribose
DNA is a much larger molecule	RNA is smaller

e Candidate A's answer suggests an understanding of the structures of the two molecules, for 1 mark, but demonstrates a lack of clarity about the facts. There is confusion as to which nucleic acid contains uracil and although the candidate knows that the sugars in the two molecules are different, this is not enough. Contrast this with the precise answers given by Candidate B (full marks). Candidate A's lack of precision is probably a result of insufficiently thorough preparation. Don't leave anything to chance — revise it all.

e Overall, Candidate A scores only 2 marks, while Candidate B scores 6. This is a quite straightforward question with no difficult concepts involved. It is just a matter of knowing the relevant biology. Examiners would expect a grade-A candidate and even a grade-C candidate to score nearly full marks on a question like this. How did you get on?

Bacterial growth

The graph shows the growth of a population of bacteria in a liquid culture at 20°C over a period of time. These bacteria are normally found in the human gut.

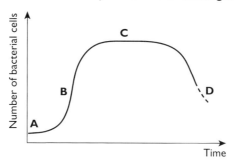

(a) Name the phases labelled **B** and **C**. (1 mark)
(b) Explain *one* reason why the number of bacterial cells shows little or no increase during the phase labelled **A**. (2 marks)
(c) State *two* reasons why the number of cells declines during the phase labelled **D**. (2 marks)
(d) On the graph, sketch the appearance of the curve you would expect if the bacteria were cultured in the same medium at 37°C. (2 marks)

Total: 7 marks

Candidates' answers to Question 2

Candidate A
(a) Stationary phase and log phase

Candidate B
(a) Phase B is the log phase; phase C is the stationary phase.

> *e* Both candidates have the right names for the two phases, but Candidate A does not make it clear which phase is which. Following the logic of the question, which asks for the names of phases B and C *in that order*, Candidate A has them the wrong way round. The mark would probably not be awarded to this candidate. You must leave no room for ambiguity in your answers. Candidate B gains the mark.

Candidate A
(b) The bacteria are adapting to new surroundings and so the cells are not dividing.

Candidate B
(b) The bacteria may have been transferred from a different medium. They may have to synthesise new enzymes before they can use the nutrients from this medium. As a result, cell division would be halted.

AS Human Biology

question

🖉 Candidate A has the general idea of adaptation to new surroundings and gains 1 of the 2 marks. Candidate B gives a more specific reason, explains it fully, and scores both marks.

Candidate A

(c) The number of cells decreases because the bacteria are dying. They are running out of nutrients and oxygen in the medium.

Candidate B

(c) This is the death phase, in which the number of cells declines because toxic waste products accumulate and inhibit metabolic pathways. Therefore, cell division is inhibited. Also, the oxygen concentration is reduced and so the bacteria are unable to respire aerobically and release the energy needed for the metabolic processes involved in cell division.

🖉 Both candidates score 2 marks. However, Candidate B has gone beyond what was required. The question asks you to state the factors, not to explain them. Although Candidate B has not lost any marks, the answer will have used up time unnecessarily. Always read the question carefully and do exactly what is required of you.

Candidate A
(d)

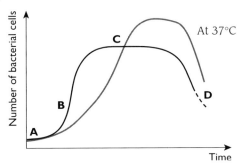

Candidate B
(d)

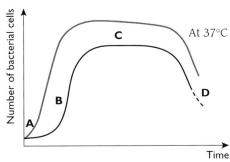

e In the stationary phase, the population of bacteria is at the maximum level that the medium can support. This will not change with temperature, although the rate of reproduction will. The stationary phase will be achieved more quickly at a higher temperature (assuming that this is not beyond the optimum for the bacterial enzymes). Neither candidate realises that the stationary phase will be at the same level, although Candidate B does realise that it will be achieved more quickly for 1 mark. Candidate A mistakenly thinks that at 37°C the log phase will be slower than at 20°C, and scores no marks.

e **Candidate A scores 3 marks and Candidate B scores 6. This is quite a straightforward question, and you might reasonably expect to be set a similar question in the unit test. Grade-C candidates should be able to score most of the marks on a question like this.**

AS Human Biology

Question 3

Mitosis and meiosis

Figure 1 shows a cell in a stage of mitosis. The cell contains just two pairs of homologous chromosomes.

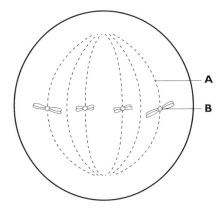

Figure 1

(a) (i) What are *homologous chromosomes*? (1 mark)
   (ii) Identify the structures labelled **A** and **B** on Figure 1. (2 marks)
   (iii) Name the stage of mitosis represented in this diagram. Give a reason for your answer. (1 mark)

Figure 2 shows the life cycle of a mammal.

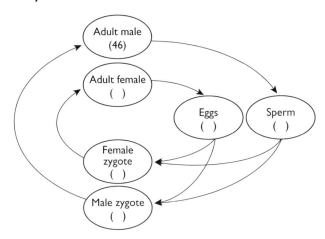

Figure 2

(b) (i) Mark on the diagram *one* stage where meiosis takes place and *one* place where mitosis takes place. (2 marks)
   (ii) Complete the empty boxes to show the number of chromosomes per cell. (1 mark)

Total: 7 marks

AQA (A) Unit 3

Candidates' answers to Question 3

Candidate A
(a) (i) They have the same genes.

Candidate B
(a) (i) A homologous pair of chromosomes is a pair of chromosomes that have the same genes along their length (although they may not have the same alleles).

> 🖉 Both candidates score the mark but, again, Candidate B has written much more than is necessary. Just answer the question, without writing part of it out again.

Candidate A
(a) (ii) **A** is the spindle; **B** is a chromosome.

Candidate B
(a) (ii) **A** is the spindle or, more accurately, one of the spindle fibres.
B is the centromere, which holds the chromatids in a chromosome together.

> 🖉 Candidate A has not looked carefully enough at label **B** which indicates, precisely, the centromere, and so scores only 1 mark. Candidate B is awarded both marks.

Candidate A
(a) (iii) Prophase

Candidate B
(a) (iii) Metaphase

> 🖉 Candidate A has not revised mitosis effectively. This is a straightforward piece of biological knowledge which any candidate who has prepared thoroughly should know. Candidate B is awarded the mark.

Candidate A
(b) (i) and (ii)

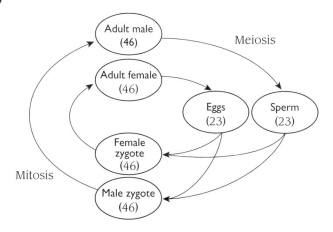

AS Human Biology

question 3

Candidate B
(b) (i) and (ii)

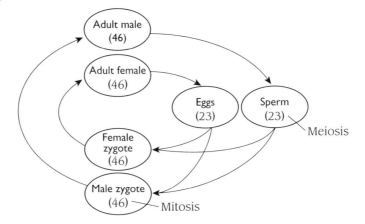

🄔 Candidate A is awarded 2 marks for part (b) (i). Candidate B also probably knows exactly where and when the two processes occur, but the labelling is unfortunate. Meiosis does not occur actually *in the sperm*, so this mark is not awarded. However, the zygote does divide by mitosis and so this mark can be awarded. **You will be expected to know that meiosis occurs *only in the formation of gametes* and that all other cell divisions are by mitosis. You should be able to relate this to a life cycle that you have not studied before.** Both candidates understand the halving of chromosomes in the gametes and the restoration of the normal (diploid) number in the zygote. Each is awarded the mark for part (b) (ii).

🄔 **Candidate A scores 5 marks for this question, while Candidate B scores 6. Most of this question is fairly straightforward biology and examiners would expect both grade-A candidates and grade-C candidates to score well. But remember, they're only easy if you know the answers — so prepare thoroughly.**

Protein synthesis

Protein synthesis takes place in the ribosomes. The code for synthesis of a particular protein is specified by a section of the **DNA** molecule and is carried to the ribosomes by mRNA.

(a) (i) What do we call a section of **DNA** that codes for a protein? (1 mark)

(ii) The **DNA** code is sometimes called a *degenerate* code. What does this mean? (2 marks)

(b) Figure 1 shows protein synthesis taking place in a ribosome.

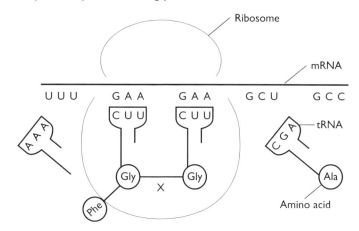

Figure 1

Key to amino acids
Phe = Phenylalanine
Gly = Glycine
Ala = Alanine

(i) Name the type of bond labelled **X**. (1 mark)
(ii) Use examples from the diagram to explain the terms *codon* and *anticodon*. (2 marks)

Total: 6 marks

■ ■ ■

Candidates' answers to Question 4

Candidate A
(a) (i) A gene

Candidate B
(a) (i) A small section of DNA that codes for a protein is called a gene.

e Both candidates know that a gene codes for a protein and each is awarded 1 mark.

Candidate A
(a) (ii) There are more codes than there are amino acids.

Candidate B
(a) (ii) Not all the triplet codes of the DNA code for amino acids. Some are stop codes.

question

> Neither candidate completely answers the question. Each is awarded **1** mark for their correct statement. If they had each included the statement made by the other or had included the idea that some amino acids have more than one code, they would have scored both marks.

Candidate A

(b) (i) Glycosidic

Candidate B

(b) (i) Glycosidic bond

> Bonds between amino acids are **peptide bonds**. Both candidates appear to have been confused by the label 'gly' in the diagram. The key makes it quite clear that this is an abbreviation for the amino acid glycine. **Look carefully at all the information you are given in a question.**

Candidate A

(b) (ii) GAA is a codon; CUU is an anticodon

Candidate B

(b) (ii) A codon is a triplet of bases on the mRNA molecule — such as GAA. An anticodon is a triplet of bases on the tRNA molecule.

> Neither candidate really makes both points here. Candidate A is awarded 1 mark as (in this instance) the examples chosen can only be codon and anticodon respectively and also *they are complementary* — but is this good luck or does the candidate really understand that codon and anticodon must be complementary? Candidate B is not awarded any marks, despite probably understanding the concepts quite well. There is no example of an anticodon and there is no indication that codon and anticodon are complementary or that they code for an amino acid. **The examiner will not assume *anything* for you: you must explain *everything*.**

> Candidate A scores 3 marks, while Candidate B scores 2. This is a topic that many candidates find difficult, as it pulls together knowledge from a number of areas. However, much of it depends only on learning the relevant facts.

Immunity

Part of our immune response involves the production of antibodies in response to specific foreign antigens. Figure 1 shows the levels of the antibody response to two injections of the same antigen.

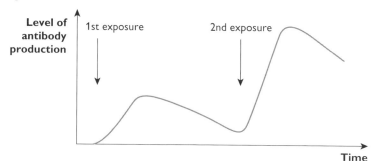

Figure 1

(a) (i) What is an antigen? (1 mark)
   (ii) Describe two differences between the responses to the first and second injections of the antigen. (2 marks)
(b) Explain how we become immune to a disease following exposure to the microorganism causing the disease. (3 marks)

Total: 6 marks

■ ■ ■

Candidates' answers to Question 5

Candidate A
(a) (i) An antigen is something that shouldn't be in your body.

Candidate B
(a) (i) An antigen is a foreign substance — often a foreign glycoprotein on the surface of a cell.

> 🄔 A spanner shouldn't be in your body — but it is not an antigen. Candidate A's answer is totally inadequate, although he or she may well have known more and, if someone had said 'be more precise', may well have done so. Candidate B clearly understands what an antigen is — although it needn't be foreign. You have your own 'set' of antigens on your cells.

Candidate A
(a) (ii) The first response produces more antibody and it drops to a lower level.

Candidate B
(a) (ii) The second response is quicker and produces more antibody than the first.

> Candidate A's answer is wrong — 0 marks. Candidate B has noticed the difference in gradient as well as the difference in amount produced, and is awarded both marks. **Look carefully at all aspects of graphs. Differences in gradient can indicate differences in rate.**

Candidate A
(b) Lymphocytes in our blood system produce antibodies which destroy the microorganisms that have infected us. Some of the lymphocytes become memory cells which stay in our body and can fight off future infections.

Candidate B
(b) When a microorganism enters our body, B lymphocytes are stimulated to produce specific antibodies against the antigens on the microorganisms. They multiply and eventually destroy all the microorganisms. Some of the lymphocytes become memory cells and remain in our bodies. If we are infected again by the same microorganism, these memory cells can destroy it quickly.

> Both candidates seem to understand what happens, but neither has supplied sufficient detail. Examiners would hope to see a mention of B lymphocytes multiplying and producing *specific* antibodies. Some of these would become memory cells and remain in our bodies and, upon re-infection, would be able to destroy the microorganisms quickly *because they are present in much larger numbers than at the initial infection*. Candidate A scores 1 mark and Candidate B 2 marks.

> **Overall, Candidate A scores just 1 mark, while Candidate B is awarded 5. Marks can be easily lost on an apparently straightforward question like this by not looking carefully enough at, and not using, all the information supplied. When you have to write more than just a line or two for an answer, read it through and ask yourself 'So what?'. If you can find an answer to the 'So what?', you probably need to include it!**

Tumours

Tumours are formed when **DNA** in a single cell undergoes a mutation that removes the natural controls on its division. As a result, repeated divisions form a clone of mutant cells. The tumour may be benign or malignant.

(a) What is a clone? (1 mark)
(b) Describe *two* differences between a benign tumour and a malignant tumour. (2 marks)
(c) A tumour in the breast cannot be felt until it has a volume of $1\,cm^3$. It takes 33 cell doublings for a single cell to produce a tumour of this size. Each cell doubling takes 30 days.
  (i) For how many years has the tumour been developing before it is detectable? Show your working. (2 marks)
  (ii) Use your answer to (i) to explain why early detection of breast cancer is essential. (2 marks)

Total: 7 marks

Candidates' answers to Question 6

Candidate A
(a) A clone is an individual that is similar to its parent.

Candidate B
(a) A clone is a group of genetically identical cells or individuals.

> 🅮 Neither candidate has given the complete answer. A clone is a group of genetically identical *individuals derived from the same parent.* Neither candidate scores the mark.

Candidate A
(b) A benign tumour is usually harmless — it is not life threatening. A malignant tumour can be fatal. Also, benign tumours cannot spread to other tissues.

Candidate B
(b) Benign tumours are usually encapsulated — having a fibrous capsule that encloses them. Malignant tumours are not encapsulated. Also, malignant tumours can spread to other tissues, whereas benign tumours do not.

> 🅮 Both candidates appreciate that malignant tumours show metastasis and benign tumours do not. However, Candidate A's answer regarding tumours being life threatening or not is both inaccurate (some benign brain tumours can be fatal) and rather vague in the context of this question. Encapsulation as explained by Candidate B is a much more appropriate response. Candidate A scores 1 mark; Candidate B scores 2 marks.

Candidate A
(c) (i) 33 doublings take 30 days each = 990 days, so the tumour has been developing for 990 days.

question

Candidate B

(c) (i) 33 × 30 = 990 days

There are 365 days in a year (assuming no leap years), so

$$\frac{990}{365} = 2.7$$

The tumour has been developing for 2.7 years.

e Candidate A did not notice that the question asks for how many *years* the tumour had been developing and, as a result, scores only 1 mark. This is awarded because the basic calculation is correct. Candidate B scores both marks. Always read the question carefully and give your answer to any numerical question in the units requested.

Candidate A

(c) (ii) The tumour has been developing for a long time and could have spread to other areas. If the tumour is detected early, it might not have spread and could be removed successfully.

Candidate B

(c) (ii) During the time the tumour has been developing, cells could have broken away and travelled in the blood or in the lymph to other tissues and organs, causing secondary cancers.

e Both candidates clearly link the lengthy development period to the possibility of metastasis. However, only Candidate A specifically addresses the point about early detection, scoring both marks. Candidate B, despite giving a more detailed answer, only scores 1 mark. Make sure that you address all the issues demanded in a question.

e **Candidate A scores 4 marks and Candidate B, 5. The question asks for some fairly straightforward points about the features of tumours and the calculation involves only basic numerical skills. You should be able to score well on this type of question.**

Malaria

Malaria is caused by a single-celled organism that is spread by the bites of the female *Anopheles* mosquito.

(a) The female mosquito bites humans to suck blood in the early evening. The malarial parasite enters liver cells where it changes form, emerges and infects red blood cells. Inside the red blood cells, the parasite divides, forming more parasites and also immature sex cells. The parasites and sex cells are released into the blood when the red blood cell bursts, which occurs in the early afternoon. The sex cells take about 28 hours to mature.

  (i) Explain how *two adaptations of the parasite's life cycle described in the paragraph above* make it difficult for the immune system to attack the parasite. (2 marks)

  (ii) Explain how the timing of the release of the immature sex cells and the biting of the female mosquito are adaptations that ensure the maximum chance of transmission of the parasite. (3 marks)

(b) Two strategies that have been used to try to eradicate malaria are:
  - draining swamps where the mosquito larvae live
  - introducing fish that feed on mosquito larvae into waterways where the larvae live

Suggest how these strategies could be effective in eradicating malaria from these areas. (2 marks)

Total: 7 marks

Candidates' answers to Question 7

Candidate A

(a) (i) The parasite lives inside the red blood cells, where the immune system cannot get at it. The sex cells are released when they are immature.

Candidate B

(a) (i) The parasite changes form and so the antigens on its surface probably change also. This means that any antibodies manufactured when the parasite first entered are now ineffective. Also, it spends a lot of time inside human cells where the immune system cannot detect it. When the blood cells burst, the parasites are exposed to the immune system, but only for a short time, so there is little time for the immune system to react.

> *e* Both candidates have the idea of the parasite being hidden from the immune system for much of its life cycle. Candidate B also understands the significance of the change of form, whereas Candidate A misses this point. Candidate B again writes far more than is necessary to answer this question and uses up valuable time. Candidate A scores 1 mark, while Candidate B gains 2.

question 7

Candidate A

(a) (ii) The sex cells are released in the afternoon and the female mosquito bites in the early evening, so the sex cells only have to last for a few hours before they can be transferred to another mosquito.

Candidate B

(a) (ii) The immature sex cells are released from the red blood cells in the early afternoon and mature 28 hours later. This means that they are mature when the female mosquitoes bite.

> Neither candidate really links all the ideas together. Candidate B links the times of biting and release of sex cells with the time taken for the sex cells to mature (and so scores 2 marks), but does not quite make the final point that 28 hours after release (in the early afternoon) it will be early evening of *the following day*. Candidate A only really links the biting and the release and scores only 1 mark.

Candidate A

(b) If the swamps are drained, then the larvae will die and so they won't be able to transmit the parasite to more people. The same will happen if you introduce fish into waterways where the larvae live.

Candidate B

(b) Both strategies will break the life cycle of the mosquito. If there are no larvae, there can be no adults and so the parasite cannot be transmitted.

> Candidate A is careless with the answer (or perhaps misses the point). The answer suggests that the *larvae* transmit the parasite, which is incorrect. Candidate A does not score. Candidate B appreciates that the life cycle will be broken and that there will be no *adults* to transmit the parasite. Candidate B scores 2 marks.

> Overall, Candidate A scores 2 marks and Candidate B scores 6. This question requires the use of analytical skills as well as recall. It is a more difficult style of question, designed to allow grade-A candidates to show their skills. However, a grade-C candidate should be able to score some marks on this type of question.

Analysing gene structure

Gel electrophoresis is often used to analyse the structure of a gene.

(a) Describe how gel electrophoresis separates fragments of DNA. (3 marks)

(b) In one investigation, a section of DNA containing 1000 base pairs was treated with two different enzymes (enzyme 1 and enzyme 2) which cut the DNA at specific base sequences. Samples of the DNA were treated with each enzyme separately and also with the enzymes in combination. The results are shown in the diagram.

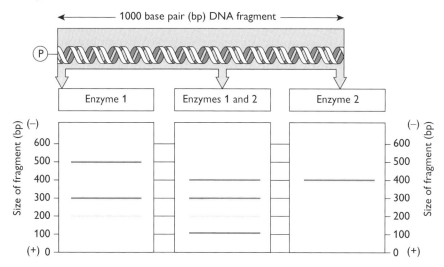

(i) What sort of enzyme would be used to cut the DNA into sections? (1 mark)

(ii) How many times did the base sequence recognised by enzyme 1 appear in the DNA? (1 mark)

(iii) On the diagram below, identify the positions where enzyme 1 and enzyme 2 cut the DNA sample. (2 marks)

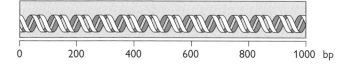

Total: 7 marks

Candidates' answers to Question 8

Candidate A

(a) Gel electrophoresis separates fragments of DNA using an electric current. Positive fragments are attracted to the negative side and negative fragments to the positive side. The distance they move tells you how big they are.

question

Candidate B

(a) Gel electrophoresis creates an electric field across a gel. The fragments of DNA are negatively charged, so they migrate towards the positive pole of the field. Larger fragments undergo more 'drag' in the gel and so do not move as far as smaller fragments. The distance moved can be used to calculate the molecular mass of the DNA fragment.

> 🖉 Both candidates seem to understand the procedure. They understand that an electric field is applied and that negatively charged particles will be attracted to the positive electrode. Candidate B describes this precisely whereas Candidate A is a little more vague. The same is true of the final point — Candidate A does not make it clear how distance travelled is related to the size of the fragment. Candidate B scores all 3 marks, whereas Candidate A scores 2 marks. An examiner would decide that one or other (but not both) of the second and third points made by Candidate A was worth a mark. Do try to be precise in your answers.

Candidate A

(b) (i) Restriction enzyme

Candidate B

(b) (i) Restriction endonuclease

> 🖉 Both candidates score the mark.

Candidate A

(b) (ii) Three times

Candidate B

(b) (ii) Twice. There are three fragments, so it must have been cut twice.

> 🖉 Candidate A makes a silly mistake here. If you take a piece of string and cut it once, you will have two pieces. Make one more cut and you will now have three pieces. The number of cuts is always one less than the number of pieces produced.

Candidate A

(b) (iii)

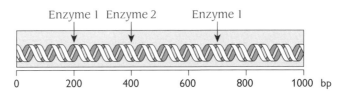

Candidate B

(b) (iii)

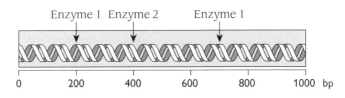

AQA(A) Unit 3

🅮 This is not an easy question. Enzyme 1 must make two cuts to produce 200 base pair (bp), 300 bp and 500 bp fragments. Enzyme 2 produces 600 bp and 400 bp fragments. Together, the 200 bp and 300 bp fragments remain, but the 500 bp fragment is split into 100 bp and 400 bp fragments. This is explained if the enzymes cut in the places shown below:

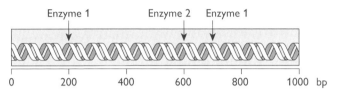

🅮 Both candidates correctly locate the cuts made by enzyme 1 (even though Candidate A wrote earlier that it would make three cuts), but not that made by enzyme 2. Both candidates score 1 mark.

🅮 **Candidate A scores 4 marks and Candidate B, 6. Although the last part of the question is difficult and targeted at grade-A candidates, the rest is relatively straightforward. Grade-C candidates should score 3 or 4 marks on a question such as this.**

AS Human Biology

Question 9

Heart disease

(a) Heart attacks often result from blockages in the coronary arteries.
   (i) Explain why a blocked coronary artery can lead to a heart attack. (3 marks)
   (ii) Explain how atherosclerosis can lead to an artery becoming blocked. (3 marks)
(b) Interruptions to the coronary circulation can sometimes be corrected using bypass surgery. The diagram shows the results of a triple bypass operation.

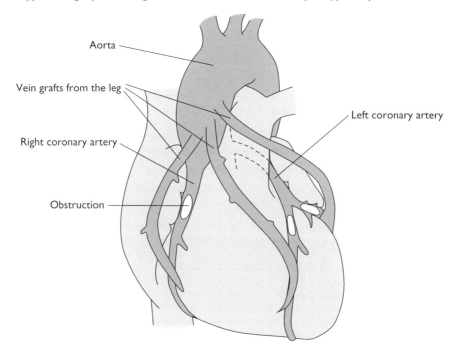

   (i) Using the example shown, explain how the triple bypass will have enabled normal heart function. (2 marks)
   (ii) Suggest why a section of a vein is used in the operation, rather than a section of an artery. (2 marks)
(c) Hypertension can also be a consequence of narrowed arteries.
   (i) Explain what is meant by hypertension. (2 marks)
   (ii) Explain how beta-blockers can be effective in the treatment of hypertension. (3 marks)

   Total: 15 marks

■ ■ ■

Candidates' answers to Question 9

Candidate A
(a) (i) A blocked artery means that the blood cannot flow through easily. This means that part of the heart cannot get blood and so is starved of oxygen and nutrients, resulting in a heart attack.

AQA (A) Unit 3

Candidate B

(a) (i) If a coronary artery becomes blocked, the area beyond the blockage receives no oxygen and so cannot respire and release energy. This makes contraction of the muscle impossible and a heart attack results.

e Candidate A understands the situation, but does not make it clear how lack of oxygen would result in a heart attack and so scores 2 of the 3 marks. Candidate B makes all three points and scores 3 marks.

Candidate A

(a) (ii) Atherosclerosis happens when cholesterol is laid down in the wall of an artery. This makes it narrower and eventually it can become totally blocked.

Candidate B

(a) (ii) Atherosclerosis occurs when there is too much saturated fat and cholesterol circulating in the plasma. These substances become deposited in the lining of arteries, which results in the artery becoming narrower. Also, the lining of the artery becomes rougher, which can trigger the clotting reaction. Any clots that form are much more likely to become stuck in a narrower artery.

e Candidate A understands the nature of the process, but once again the wording is a little imprecise. By 'wall of the artery', Candidate A might mean the muscular/elastic regions of the artery wall, which is wrong. However, Candidate A scores 2 marks for correctly identifying cholesterol and describing the narrowing that results. Candidate B gives a much more detailed answer and easily scores all 3 marks.

Candidate A

(b) (i) The vein grafts allow blood to bypass the blockages in the coronary arteries and so the heart can beat normally again.

Candidate B

(b) (i) Because blood bypasses the blockages, all regions of the heart can now receive oxygenated blood again and can respire aerobically, releasing the energy needed for muscle contraction.

e Yet again, Candidate A does not supply the necessary detail and so scores just 1 mark. The answer does not explain in sufficient detail *why* the heart can beat again. Candidate B provides a much more precise answer, for both marks. It is essential to look at the mark allocation and decide how much you need to write. If you think you need to write more than you have already written but are struggling to come up with anything, try asking yourself 'so what?' about the answer you have produced — it might just jog your memory into thinking of further detail.

Candidate A

(b) (ii) A vein is larger than an artery and so is easier to manipulate in surgery.

Candidate B

(b) (ii) Veins have a larger internal diameter than arteries.

question 9

> *e* Candidate A completely misses the point. The overall size of the vein would be chosen to match that of the coronary arteries to which it must be attached. Candidate B appreciates that it is the internal diameter that is important, for 1 mark, but fails to explain why — a larger internal diameter will give less resistance to blood flow.

Candidate A

(c) (i) Hypertension is raised blood pressure. This can be caused by stress, or in some cases, by too much salt in the diet.

Candidate B

(c) (i) Hypertension is sustained high blood pressure. It is a major risk factor in coronary heart disease.

> *e* Hypertension is correctly described by Candidate B, for 2 marks. Candidate A scores only 1 mark because the answer does not make clear that the condition is ongoing.

Candidate A

(c) (ii) Beta-blockers bind to receptors in the heart and reduce the rate and force of the contractions.

Candidate B

(c) (ii) Beta-blockers have a shape that is similar to adrenaline and noradrenaline and so bind to beta receptors in the heart muscle. Noradrenaline and adrenaline are prevented from binding, so the force and rate of contractions is reduced, bringing down blood pressure.

> *e* Candidate A does not say which receptors are involved or why the beta-blockers can bind. Also, there is no explanation of how the force and rate of heart contractions are reduced. Candidate A scores 1 mark. Candidate B scores all 3 marks.

> *e* **Overall, Candidate A scores 7 marks, while Candidate B scores 14.** This longer question is fairly straightforward, demanding mainly factual recall of basic biology relating to coronary heart disease. If you have revised thoroughly, you should score well on this type of question. You cannot afford to lose marks in the way that Candidate A does, i.e. by not supplying sufficient detail.

Question 10

Tuberculosis

(a) Tuberculosis was once so common in Europe that it was responsible for one in four deaths. This epidemic lasted from 1500 until early this century and was known as the 'great white plague'. The graph shows the death rate from TB in England and Wales from 1840 to the present time.

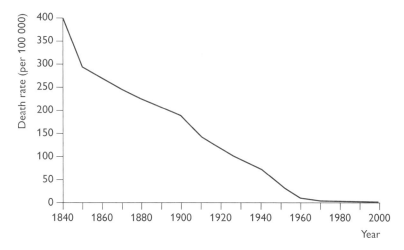

(i) List the criteria that Robert Koch used to show that *Mycobacterium tuberculosis* is the organism causing TB. (3 marks)

(ii) Describe how *M. tuberculosis* infects healthy people. (2 marks)

(iii) Suggest an explanation for the decrease in deaths from TB from 1840 to 1950. (1 mark)

(b) The structure of the cell wall of *M. tuberculosis* is complex and its synthesis involves enzymes and metabolic pathways unique to *Mycobacterium*. Use this information, and your knowledge of the action of penicillin, to suggest why most drugs developed for treatment of TB target the synthesis of the cell wall. (3 marks)

(c) Immunity to TB is almost totally cell-mediated. Because of this, AIDS sufferers are particularly at risk of contracting TB.

(i) Explain what is meant by 'cell-mediated immunity'. (2 marks)

(ii) Explain why AIDS sufferers are particularly at risk of contracting TB. (2 marks)

(d) A new vaccine against TB is being developed. This involves injecting DNA that codes for the antigens on the surface *Mycobacterium* cells. Suggest how such a vaccine might be effective. (2 marks)

Total: 15 marks

■ ■ ■

Candidates' answers to Question 10

Candidate A

(a) (i) Robert Koch found that all the people with TB had the same bacteria in their lungs and they all showed the same symptoms.

question

Candidate B

(a) (i) Robert Koch used what are now called 'Koch's postulates'. He showed that people with the disease always had the same bacteria in them and people who were healthy didn't have these bacteria. He showed that he could take the bacteria from infected people, grow them in culture and then take bacteria from this culture to infect other healthy people.

> *e* Candidate A does not recognise that the question is about Koch's postulates. However, the answer does score 1 mark (just) for the idea that the bacterium is always associated with the disease, even though the point that healthy people do not contain the bacterium is not made. Candidate B gives a fairly comprehensive answer, for 3 marks.

Candidate A

(a) (ii) The bacterium is spread by droplet infection. Once in the lungs it multiplies to form an infection focus. Here, toxins from the bacteria damage lung tissue.

Candidate B

(a) (ii) The bacterium is spread through the air in coughs and sneezes. It enters the lungs where it multiplies and causes damage to lung tissue as the bacteria release toxins.

> *e* Both candidates produce similar answers and score 2 marks.

Candidate A

(a) (iii) The living conditions in England and Wales improved during that period.

Candidate B

(a) (iii) Improved living conditions and better nutrition resulted in immune systems becoming better developed. Also, general hygiene improved over the period.

> *e* Both candidates know that living conditions have an enormous impact on the effect of disease. Both score the mark.

Candidate A

(b) If the cell wall structure is complex, a drug that can interfere with the manufacture of just one part of it will mean that the bacteria cannot synthesise the cell wall properly and so cannot multiply properly.

Candidate B

(b) A complex cell wall means that there are many components and a number of drugs are possible, each targeting a different component or enzyme system. Lack of just one component will result in a weakened cell wall. This could result in death of the bacterium by osmotic lysis — the rupture of the cell wall as water enters by osmosis. Ordinarily, a strong cell wall would resist this entry and the cell would not burst.

> *e* Candidate A proposes a bacteriostatic mode of action for the drugs. However, although the answer scores 2 marks, it cannot be awarded all the marks as it does

AQA (A) Unit 3

not take into account 'your knowledge of the mode of action of penicillin', as requested in the question. Candidate B's answer proposes a bactericidal mode of action which does take account of osmotic lysis, and so scores 3 marks.

Candidate A

(c) (i) Cell-mediated immunity results from the action of the T-lymphocytes. It does not involve any plasma cells or antibodies being produced.

Candidate B

(c) (i) Cell-mediated immunity results when cytotoxic T-lymphocytes multiply and kill invading pathogens directly, without the need for antibody production.

e Both candidates clearly understand the concept and score 2 marks.

Candidate A

(c) (ii) AIDS sufferers are at risk of contracting TB because AIDS weakens their immune system and they cannot fight off the AIDS virus.

Candidate B

(c) (ii) AIDS sufferers are particularly at risk of contracting TB as HIV infects the T-lymphocytes that are involved in the cell-mediated immunity to TB. Because of this, the incidence of TB worldwide is on the increase again, particularly in areas where AIDS is common.

e Candidate A's answer is in part imprecise and in part confused, and scores no marks. More detail is needed than just saying that AIDS weakens the immune system. Also, the candidate is confused as to which microorganism would be affected by the immune system. Candidate B clearly links the destruction of T-lymphocytes by HIV to the fact that these are the cells that give immunity to TB, for 2 marks.

Candidate A

(d) The lymphocytes would produce antibodies against the DNA and so the person would become immune.

Candidate B

(d) The DNA would stimulate the production of antigens, which would stimulate an immune reaction.

e Candidate A incorrectly suggests that immune responses are brought about by DNA. Candidate B does not make the point clearly that the DNA would cause the antigens to be made by coding for their production in human cells. Neither candidate scores any marks.

e **Candidate A scores 8 marks and Candidate B scores 13. This is a fairly typical question about disease and its treatment. It tests recall as well as the ability to interpret information and apply concepts to new situations. A well-prepared candidate should score over half marks on a question such as this.**

question

Examiner's overview

There are 83 marks available in these ten questions. The unit test will be slightly shorter, with a total of 75 marks, although there will still be two long questions, each worth 15 marks, at the end of the test.

Candidate B scores 69 of the 83 marks available (equivalent to 62 out of 75 marks); this is the work of a very good grade-A candidate. You could be awarded a grade A with fewer marks than this.

Candidate A scores 39 of the 83 marks available (equivalent to 35 out of 75 marks), which is probably not quite enough to be awarded a grade C.

This candidate does not perform evenly throughout the paper. Some questions are answered quite well while others are answered poorly. This suggests gaps in knowledge and understanding, probably due to insufficient preparation. This candidate could have scored another 7 marks just by being more careful:

- In question 2(a), there is ambiguity in the answer. The candidate did not make clear which name refers to which phase on the growth curve (1 mark lost).
- In question 3(a)(ii), the candidate did not look at label line B carefully enough (1 mark lost).
- In question 5(a)(ii), the candidate described the first immune response as higher; the second response produces more antibody and the candidate should not have missed this (1 mark lost).
- In question 6(c)(i), the candidate left the units of the answer as days instead of converting to years (1 mark lost).
- In question 7(b), the candidate confused mosquito larvae with adults (2 marks lost).
- In question 8(b)(ii), the candidate did not equate correctly the number of cuts made by an enzyme with the number of pieces formed (1 mark lost).

These extra 7 marks would have ensured the candidate a grade C and may have just gained a grade B. However, the story does not end here. There are many occasions throughout the paper where the candidate clearly understands the principles involved but simply does not provide the detail needed in the answers. By supplying the necessary detail, Candidate A could have scored another 5 marks in Question 9 and another 3 marks in Question 10, which would have turned this performance into a good grade B and possibly a grade A!

The message is simple: prepare thoroughly and be careful in the examination.